I0145822

ORGANIZATIONAL TRANSPARENCY, EMPLOYEE

PERCEPTIONS, AND EMPLOYEE MORALE: A

CORRELATIONAL STUDY

by

Harroll J. Ingram, JR

A Dissertation Presented in Partial Fulfillment

of the Requirements for the Degree

Doctor of Management in Organizational Leadership

UNIVERSITY OF PHOENIX

January 2009

Copyright © 2009, 2019 by Harroll J. Ingram

All rights reserved. No part of this book may be reproduced or used in any manner, except for the use of quotations, without written permission of the copyright owner.

Published (2019) by H.I. Enterprises, www.hienterprises.us

ISBN 978-1-7341831-0-8

ABSTRACT

Surveys released in 2007 reported that only 57% of Homeland Security employees were satisfied with their jobs and only 67.5% of employees across the Federal Government rated their jobs as satisfactory (Losey, 2007). This quantitative correlational study addressed the problem that leaders' use of hiring and promoting practices may influence employee perceptions of fairness and morale. The study, conducted with civilian employees at an Army organization in Florida, examined the relationships between transparency, employee perceptions of fairness, and employee morale. The results of the current research study revealed a strong correlation between hiring and promoting practices and employee perceptions of fairness/employee morale. The findings indicated that increased transparency in hiring and promoting processes can enhance employee perceptions of fairness and morale.

DEDICATION

The current research study is dedicated to my immediate and extended family members who gave unwavering support throughout my doctoral studies. My mother inspired me since my youth to strive to be the best I could be without causing harm to others. Her inspiration was a valuable part of this research study. My sister provided the driving force that propelled me during difficult times throughout this dissertation journey. I am eternally grateful for the support my family and friends offered over the last four years.

I dedicate this research project to my employer and the United States (U.S.) taxpayers. My employer assisted in financing this endeavor and served as an impetus to the recognition of the problem areas researched in the current research study. The latitude offered by my employer facilitated my ability to illuminate leadership areas that should become focus areas in an effort to better serve the U.S. taxpayers. The concerns raised by the taxpayers spurred this quest to research and seek answers that might lead to changes that would improve the lives of the people

of this nation called the United States of America.

ACKNOWLEDGEMENTS

I extend great appreciation and gratitude to my mentor and committee chairperson, Dr. Linda de Charon, for her guidance and patience throughout the dissertation process. Dr. de Charon's ability to correct, diplomatically, my stubbornly-held inaccurate positions on how to complete this project made the journey very educational and was directly responsible for the successful completion of this dissertation. I would also like to express my sincerest appreciation to committee members Dr. William Shriner and Dr. Marilyn Griffin for their continued support and contributions to this dissertation process. Their actions helped me realize this significant goal.

I am grateful to my friends and past coworkers Mr. Reginald Smith for preparing me mentally to move forward, Mr. Kevin Senior for urging me to pursue a doctoral degree, Ms. Una Brown for positively influencing my decision to enroll at the University of Phoenix, and Ms. Clara Moss for patiently supporting me through the late nights filled with reading, writing, and group project meetings. I am thankful to my baccalaureate classmate and hometown friend Dr. Eric Parrish for helping me realize financing options for doctoral studies and pushing me to start the doctoral journey. I am also thankful to my doctoral colleague Ms. Twana Arnold for spurring me into action on several occasions to stimulate progress on completing this dissertation.

Most important, none of this would have been possible without the love and patience of my mother Dr. Gloria J. Ingram. My mother is a lifelong educator. Because Dr. Ingram serves, unknowingly, as our role model, my sister and I chose to follow her lead and earn a doctoral degree.

TABLE OF CONTENTS

LIST OF TABLES

LIST OF FIGURES

CHAPTER 1: INTRODUCTION

A survey released in January, 2007 reported that only 57%
of the Homeland Security employees were satisfied with their jobs.
Only 67.5%, across the Federal Government, gave satisfactory
reports concerning their jobs (Losey, 2007). Scholars seek causes
of less than optimal job-satisfaction reports and ways to improve
the results of those reports. Research has shown that positive
perceptions concerning organizational promotion systems have a
positive effect on employee attitudes (Bagdadli, Roberson, &
Paoletti, 2006). Researchers, regarding nurses at three hospitals,
reported personal attitude as one of three factors that represent
organizational morale (Anonymous, 2008).

Morale is defined as the degree to which an employee
exhibits a positive or motivated psychological state and has
positive feelings concerning his or her work and work environment
(McFadzean & McFadzean, 2005; McKnight, Ahmad, &
Schroeder, 2001). Morale and positive employee perceptions may
be influenced by transparency. Transparency ensures all parties

have access to information about the organization (Lainhart, 2000). When transparency is absent, views become distorted and misunderstood (Richardson, 2004). The current research study examined the relationships between organizational transparency, employee perceptions, and morale at an Army organization in Florida.

Chapter one will propose research relative to the relationship between employee morale and employee perceptions of fairness concerning organizational hiring and promoting practices. Research relative to the relationship between organizational transparency and employee perceptions of fairness will be presented in this chapter. Chapter one will provide background information concerning employee perceptions of fairness and how those perceptions have been found to affect commitment, morale, and organizational effectiveness.

Negative repercussions of making unfair hiring and promoting decisions will be discussed in this chapter. A statement of the problem indicating a need for the current research study will be expressed and the research design used to examine the research problem will be introduced. The purpose and significance of the current research study will be revealed along with the targeted general population, study variables, research questions, and hypotheses. Chapter one will propose research that examines the relationships between organizational transparency, employee perceptions, and morale and that identifies if employee perceptions that leaders are using unfair hiring and promoting practices degrades employee morale at an Army organization in Florida.

Background of the Problem

Bagdadli et al. (2006) found that the perception of fairness in promotion systems positively influences employee attitudes and leads to increases in organizational commitment. Perceptions of organizational fairness guide employees' level of trust in the organization (Rajnandini, Schriesheim, & Williams, 1999). Colquitt, Conlon, Wesson, Porter, and Ng (2001) found that employees who perceive unfairness within the organization (i.e., feel they have been treated unfairly by their organizational leaders) are likely to respond with reduced commitment and job performance. Makawatsakul and Kleiner (2003) found that employee uncertainty and fear can lead to a decline in morale and organizational productivity.

Messer and White (2006) found that perceived fairness significantly affects organizational effectiveness and performance that supports the social and psychological organizational environment. Organizational members should be nurtured in a way that makes them feel productive and instrumental in organizational effectiveness and success (Gummer, 2001). Organizational productivity was found to be affected by unfair hiring practices such as preferential hiring (Singer & Singer, 1999). Hiring practices and procedural justice in promotion decisions have been found to affect employee commitment (Lemons & Jones, 2001).

Past research examined the effects of procedural and distributive justice on employee commitment and productivity. Researchers have conducted studies describing a relationship between employee perceptions of fairness and organizational commitment. Research spanning the last 18 years, relative to procedural and distributive justice and the relationship between fairness perceptions and commitment, was instrumental in the current research study.

Research, including an assessment of reactions to different types of selection techniques used by employers, supported the notion that fairness perceptions concerning employee selection processes can influence positive and negative reactions. Fairness perceptions can foster positive reactions when fairness is perceived or negative reactions when unfairness is perceived. Fodchuk and Sidebotham (2005) mentioned the importance of understanding that negative reactions due to perceived unfairness may precipitate a discrimination lawsuit against an employer.

Sharing information with employees is believed to promote fairness perceptions. Truxillo, Bauer, Paronto, and Campion (2002) conducted a study where the level of information given to candidates was manipulated to ascertain relative procedural justice perceptions. The results of the study conducted by Truxillo et al. (2002) supported the hypothesis that applicants who received more information had significantly higher perceptions of procedural justice than those who received less information.

Problem Statement

Kaplan and Ferris (2001) found that employee perceptions of fairness are influenced by the employee-selection processes used by leadership (e.g., leadership's use of position-related criteria to make promotion decisions). Promotion decisions made without using position-related criteria are sometimes viewed as unfair (Sashkin & Williams, 1990). The general problem on which the current research study focuses is that organizational leaders who do not employ transparency might be perceived as employing unfair organizational practices.

The specific problem is that employee perceptions of organizational leadership using unfair processes may degrade employee morale (Makawatsakul & Kleiner, 2003). Perceptions of unfair hiring and promoting practices might cause employees to become dissatisfied, unmotivated, uncommitted, and disloyal to the organization (Abdalla, Maghrabi, & Raggad, 1995). Lack of organizational openness, or transparency, regarding hiring and promoting practices may be viewed as unfair and may negatively impact employee morale.

This quantitative correlational study examined the relationship between employee perceptions of fairness concerning organizational hiring and promoting practices and employee morale. The study considered the relationship between organizational transparency and employee perceptions of fairness. A Likert-type survey instrument was validated and used to gather perception data, relative to the fairness of hiring and promoting practices; transparency; and morale, from civilians employed at an Army organization in Florida.

Purpose

The purpose of this quantitative correlational study was to examine the relationships between two dependent variables (i.e., the relationship between employee morale and employee perceptions of fairness concerning organizational hiring and promoting practices) and one independent variable (i.e.,

transparency). A web-based five-point Likert-type survey

instrument was used to gather employee perception data relative to

the dependent and independent variables. A sample of the 592

civilians employed at an Army organization in Florida was

surveyed to gather perception data that was analyzed to test the

hypotheses and answer the research questions of the current

research study.

Significance of the Study

The results of the current research study could be

significant to employers if the results of the current research study

support the hypothesis that a correlation exists between employee

perceptions of fairness concerning leaders' use of hiring and

promoting practices and employee morale. U.S. citizens might

recognize a need to influence organizations to ensure their leaders

foster perceptions of fairness to nurture morale and possibly

decrease the number of employee lawsuits. Employee lawsuits

against the U.S. Government are financed via taxes that are paid by U.S. citizens. One lawsuit resulted in the U.S. government agreeing to pay $508 million to 1,100 women who claimed two federal agencies denied them jobs because of their gender ("Government settles decades-old discrimination case", 2000.) Because the Government relies on tax revenue to settle many discrimination cases, the results of the current research study could lead to a reduction in taxpayer burdens.

The results of this research may influence leadership thoughts concerning the benefits of employing transparency in organizations. The study results may support the growing belief that increased information disclosure in organizations is necessary for fostering trust and morale. Practicing transparency in decision making is a positive approach to leadership (Trevino, Hartman, & Brown, 2000) and fosters open communication and trust (Stephenson, 2004). Future studies examining transparency in Federal agencies and how transparency relates to employee morale may be influenced by this research.

Significance of the Study to Leaders

The current research study could be significant to leaders if the results of the current research study support the hypothesis that increased organizational transparency fosters employee perceptions of fairness. Leaders could use the results of the current research study to justify sharing procedural information with employees to foster transparency, facilitate employee perceptions of fairness, and help the organization avoid charges of discrimination by employees. Grinberg and Nazer (2008) reported that in 2007 the U.S. Equal Employment Opportunity Commission (EEOC) received the highest volume of private sector discrimination charges since 2002. Grinberg and Nazer (2008) quoted the EEOC Chairman as saying:

Corporate America needs to do a better job of proactively preventing discrimination and addressing complaints promptly and effectively. . . . To ensure that equality of opportunity becomes a reality in the 21st century workplace, employers need to place a premium on fostering inclusive and discrimination-free work environments for all individuals (para. 1)

The results of this research could increase leadership's understanding of the relationship between organizational transparency, employee perceptions, and morale. The Federal Funding Accountability and Transparency Act of 2006 was passed by congress and signed into law to increase organizational information disclosures (Koppel, Barrett, & Tatton, 2006). The results of this research may support existing literature denoting the importance of employing transparency in Federal and private sector organizations.

Nature of the Study

This quantitative study examined the relationship between organizational transparency, employee perceptions of fairness, and employee morale using a Likert-type survey instrument. Quantitative research allows the researcher to use instruments that collect numeric data from respondents, analyze the data using statistics, and conduct "the inquiry in an unbiased, objective manner" (Creswell, 2005, p. 39). A quantitative approach was chosen for this research study because quantitative approaches are appropriate for studies where investigations are designed to capture statistically-identified factors that might influence the need for interventions (Limas, 2005; Sheldon, 2007). Creswell noted that quantitative research studies can describe correlations (2002) and facilitate problem-solving research between measurable variables, while qualitative methods focus on unknown variables (Creswell, 2005). This correlational study does not attempt to prove causation.

Qualitative approaches are used in exploratory research employing data gathering methods such as in-depth interviews (Sheldon, 2007). Because of the controversial nature of the subject of the current research study, complete anonymity likely increased the response-rate and fostered more reliable responses to the survey items. The quantitative approach used for this research facilitated accomplishing the goals of the current research study (e.g., examining relationships to reveal correlations). To foster anonymity and an increased survey response-rate, a qualitative research approach (e.g., conducting face-to-face interviews to explore and understand a central phenomenon) was not chosen for the current research study.

The population for the current research study was composed of 592 Army civilian employees. The civilian employees, part of a highly trained workforce, were required to complete 40 hours of annual training designed to prepare employees for future promotions. Many of the employees of the Army organization applied for upward mobility positions (i.e., promotions) within the organization.

Research Questions

Research questions allow researchers to fully examine a topic. Determining the research question(s) is an important step in the research process because the questions narrow the research objective and purpose to specific areas on which the researcher can focus his or her studies (Creswell, 2005). The research questions of quantitative studies are designed to show possible relationships between variables (Creswell, 2005). The following two research questions were posed for this quantitative study:

R1: What is the relationship between employee morale and the fairness perceptions of employees concerning organizational hiring and promoting practices?

R2: What is the relationship between increased transparency in an organization's hiring and promoting practices and employee fairness perceptions?

Hypotheses

Bagdadli et al. (2006) found that the perception of fairness in promotion systems positively influences employee attitudes. Based on research relative to morale, perception of fairness, and transparency, the following sets of hypotheses were advanced for the current research study.

H1 null (H1$_0$): A correlation does not exist between employee perceptions of fairness concerning leaders' use of fair hiring and promoting practices and employee morale.

H1 alternate: A correlation exists between employee perceptions of fairness concerning leaders' use of fair hiring and promoting practices and employee morale.

H2 null (H2$_0$): Increased transparency in organizational hiring and promoting processes does not greatly foster employee perceptions of fairness.

H2 alternate: Increased transparency in organizational hiring and promoting processes greatly fosters employee perceptions of fairness.

Research Variables

Quantitative research designs may include several types of variables such as independent, dependent, and moderating variables. The current research study employed an independent variable and two dependent variables. An independent variable is manipulated to determine its relationship with a dependent variable. The independent variable in the current research study was organizational transparency and the dependent variables were employee morale, and employee perceptions of fairness concerning leaders' use of hiring and promoting practices.

Theoretical Framework

A major framework of the current research study concerned a germinal concept known as equity theory. Equity theory explains a balancing relationship between investments and outcomes. Adams (1965), in germinal writings concerning equity theory, postulated that the degree to which an exchange is equitable is determined by comparing the investment (attention, skills, and efforts) and outcome (status, appreciation, and pay) ratio of one person to the investment and outcome ratio of another person. Equity theory evolved into an often discussed theory known as procedural justice.

Procedural justice was one of the theories composing the theoretical framework of the current research study. Gilliland (1993) suggested that procedural justice is related to how closely leaders adhere to procedural rules. The remaining two theories that completed the three-part theoretical framework of the current research study were employee morale and transparency.

The results of the current research study provides scrutiny of research supporting the notion that employee perceptions of fairness are influenced by leadership's employee-selection methods (Kaplan & Ferris, 2001) and notions that employee perceptions of leadership's use of unfair hiring and promoting processes affect employee morale (Makawatsakul & Kleiner, 2003). Research results over the last seven years provide a close examination of research-backed notions that organizations benefit from practicing transparency when revealing policies and procedures (Garten, 2002) and that employing transparency positively affects employee morale (Limas, 2005). The results of the current research study augment major theories concerning procedural justice, morale, and transparency while offering insight into the research problem that was studied.

Procedural justice deals with the treatment people receive from their leaders and other decision makers. Treatment relative to honesty, courtesy, timely feedback, respect for rights, and adequate explanation of decisions (Furnham & Petrides, 2006). Procedural justice is rooted in another traditional model called organizational justice. Organizational justice focuses on fairness perceptions of decision outcomes (Lambert et al., 2005). Organizational justice is believed by some scholars to be three dimensional i.e., consist of procedural justice, distributive justice, and interactional justice (Peelle, 2007). Other scholars recognize a strong correlation between procedural justice and interactional justice (Cropanzano, Prehar, & Chen, 2002).

Employee morale is concerned with an individual's mental and emotional condition and is believed to be related to enthusiasm, confidence, and loyalty (Haun, Vivero, Leach, & Liuzza, 2002). Early research into employee morale revolved around finding a way to decrease absenteeism. Eventually, scholars realized the concept of employee morale was more dynamic and depended on employee satisfaction and goal attainment (Baehr & Renck, 1958).

Research over the last two decades added controversy to the notion that fairness perceptions foster morale. Sashkin and Williams (1990) found that morale might not be nurtured in an environment where an employee believes leadership's actions are fair. Although an employee might feel leadership's actions are fair, those actions might leave the employee unhappy and negatively affect the employee's morale (Sashkin & Williams, 1990).

Transparency ensures all parties have access to information about the organization (Lainhart, 2000). The importance of employing transparency surfaced as a result of increased competition between multinational companies which led to increased corruption (Denker, 2007). Federal Government responsibilities, following the Great Depression, grew along with increased influence of administrative agencies that applied regulations without close legislative review. The Government eventually passed laws such as the U.S. Administrative Procedure Act of 1946 and the Freedom of Information Act of 1966 to give more access to department and agency information (Roberts, 2004).

Practicing transparency has facilitated internal and external organizational information exchanges. Network sharing and the use of open systems approaches promote the use of transparency and can be beneficial to organizations (Banff Executive Leadership Inc., 2003). Practicing transparency will foster the sharing of information concerning employee selection processes (i.e., will ensure employees understand the processes and criteria used to hire and promote employees). When transparency is absent, views become distorted and misunderstood (Richardson, 2004). Promoting transparency facilitates a work environment where good moral examples are set (Somiah, 2006).

Definitions

The following definitions are provided for key terms that are subject to a wide variety of interpretations.

Distributive justice: Fair and equitable distribution of resources such as salary and bonuses (Blader & Tyler, 2003).

Morale: The reflection of an individual's psychological well-being, confidence, and sense of purpose. High morale can be indicated by an individual's positive attitude, optimism, and cooperative behavior ("Center for Creative Leadership", 1993).

Interactional justice: Relates to the quality of treatment received from leadership (Tyler & Bies, 1990).

Procedural justice: Conducting procedures honestly and with consideration for the views and rights of all concerned (Furnham & Petrides, 2006).

Prospect theory: A concept that explains the value of considering expectancies, when finalizing decisions between options, and identifies the risk tolerance of the decision maker

(Miceli, n.d.).

Transparency: Remaining free from pretense or deceit

(Banff Executive Leadership Inc., 2003) when revealing internal

controls used to govern an organization ("Baldrige glossary",

2007).

Assumptions

Employees may feel the need to protect their organization

by not offering responses to the survey items that could potentially

be viewed as negative. The Army periodically conducts surveys to

gauge employee perceptions and emphasizes the importance of

facilitating organizational improvements by answering survey

questions truthfully. The reliability of the results of the current

research study is predicated on the assumption that integrity

instilled in the Army employees will lead the respondents to

provide truthful responses to the survey items.

Respondents might hold a fear of being identified as responsible for giving negative responses to certain survey items. The current research study showed consideration for the confidentiality of respondents and included measures to encourage openness and facilitate anonymity. The assumption was made that respondents will recognize the confidentiality and anonymity considerations of the methodology of the current research study and respond to survey items without allowing fears to interfere with the openness of their responses.

Scope

The current research study concerned the fairness perceptions of civilian employees relative to the hiring and promoting practices employed at an Army organization located in Florida. Whether or not a correlation exists between the employees' fairness perceptions and their level of morale was examined in the current research study. The current research study considered the relationship between organizational transparency and employee perceptions of fairness. The civilian population of the Army organization used for the current research study consisted of 592 employees. A random sample of the population was solicited to participate in the current research study and the data was collected using a Likert-type survey instrument.

Limitations

The current research study was limited by the number of employees who responded to the web-based five-point Likert-type survey instrument containing 10 items seeking employee perception data. The researcher received the number of responses necessary to render the results of the current research study statistically representative of the population. The current research study may have been limited by the number of honest responses given relative to the survey items. A lack of honest responses to the survey items can render the study unreliable. Despite any relationship findings between organizational transparency, employee perceptions of fairness concerning leaders' hiring and promoting practices, and employee morale, the conclusions drawn from the findings are limited because those conclusions are based on perceptions.

The civilians surveyed for this research work within a military construct that may not be keeping pace with corporate organizations in reference to adopting organizational improvement concepts such as organizational justice and organizational transparency. The results of this research, consequently, may not be generalizable to civilian populations at all organizations. The generalizability of the results of the current research study may need to be confined to civilian populations working in military-led organizations within the public sector.

Delimitations

Delimitations are researcher-controlled study limits. The current research study contained two delimitations. First, the Likert-type survey instrument used was limited to statements requesting closed-ended responses. Closed-ended responses do not allow the opportunity for follow-on inquiries designed to develop a deeper understanding of the reasons behind the responses. Responses to closed-ended statements are final, limit response value-assessments, and do not foster opportunities to identify and discard survey responses that are based on irrelevant perceptions (e.g., personal vendettas).

The second delimitation concerns the size of the population and survey sample. The sample chosen for the current research study was limited to a small civilian population employed at an Army organization located in Florida. The decision to not include the organization's military population in the sample may have negatively affected the generalizability of the results of the current research study.

The aforementioned delimitations may be considered as shortcomings of the current research study, but the delimitations are not believed to have a negative impact on the generalizability of the results of the study. The generalizable results of the current research study likely represent the perspectives of the population of the Army organization. If the results of researching a sample group can be generalized to a population, the results are perceived as more useful and more important (Stoddart, 2004).

Chapter Summary

Chapter one addressed the problem that employee perceptions of fairness, influenced by leaders' use of hiring and promoting practices, may decrease morale (Makawatsakul & Kleiner, 2003). The current research study examined the extent of the relationship between employee perceptions of fairness concerning organizational hiring and promoting practices and employee morale at an Army organization in Florida. The current research study examined the relationship between organizational transparency and employee perceptions of fairness concerning organizational hiring and promoting practices. The study used a Likert-type survey instrument, validated using a pilot study, to gather information from civilians employed at the Army organization.

Two research questions were posed in chapter one. The first research question inquired about the relationship between employee morale and the fairness perceptions of employees concerning organizational hiring and promoting practices and the second research question inquired about the relationship between increased transparency in an organization's hiring and promoting practices and employee fairness perceptions. Two sets of hypotheses were expressed in the chapter to assist in answering the research questions.

Chapter one described the methodology and theoretical body of knowledge that was used to answer the research questions. The methodology that was used for the current research study was a quantitative correlational method. The theoretical body of knowledge that guided the current research study was described as a three-part theoretical framework composed of the procedural justice, employee morale, and transparency theories.

Chapter two will build on the theoretical framework mentioned in chapter one. The literature reviewed will cover procedural justice, employee morale, and transparency. The

objective of chapter two will be to synthesize how researchers

have used theories and research designs to describe the effects of

employee perceptions of fairness on employee morale and how

increased transparency affects employee perceptions of fairness.

CHAPTER 2: REVIEW OF THE LITERATURE

The research study examined the relationship between employee perceptions of fairness, relative to organizational hiring and promoting practices, and employee morale. The relationship between organizational transparency and employee perceptions of fairness was also analyzed in the current research study. Chapter two will examine historical, current, and gaps in literature concerning relationships between organizational transparency, employee perceptions, and employee morale. Chapter two will review the literature pertinent to the research questions and hypotheses and the relevant theories. The overarching theoretical body of knowledge that will be used will be a tripartite framework encompassing procedural justice, employee morale, and transparency.

Title Searches, Articles, Research Documents, Journals

The literature reviewed was obtained from several sources including peer-reviewed articles in scholarly journals and other germinal, key, and current sources. The literature reviewed

primarily focused on the broad subjects of procedural justice,

employee morale, and transparency and subsequently concentrated

on more specific areas such as procedural justice in hiring and

promoting practices, organizational communications concerning

hiring and promoting practices, and fairness perceptions as related

to maintaining employee satisfaction. Pertinent key words and

phrases were searched to assist in the literature review.

The literature search covered both electronic and print

academic sources. Electronic sources including the ProQuest

Research Database, EBSCOhost Library, Emerald Research

Database, and Online Encyclopedias were searched. Others

sources included Transparency International publications and

materials from other credible sources (e.g., Federal Times online,

U.S. Employment Opportunity Commission articles, and Institute

For Global Ethics Website). The print sources included articles and

books not available in the electronic databases, but authored by

renowned researchers. An examination of the literature reviewed exposed areas that had not been researched and gaps in literature which are discussed within the Current Findings section below.

Historical Overview

Research concerning employee perceptions has been progressing for over 40 years. Employee perceptions relative to organizational fairness is believed to be predicated on employee investment and outcome comparisons. Scholars have examined theories concerning relationships between employee investments and outcomes such as Adams equity theory (1965) and what constitutes fair processes in organizations (Blader & Tyler, 2003).

Employee morale research was prompted by early studies related to employee needs (Baehr & Renck, 1958). The satisfaction of employee needs was believed to prompt drive and influence morale. Morale, according to Baehr and Renck (1958), is structured by basic factors related to the employee's relationship with management and how the organization conveys information. Researchers postulated that organizations would benefit from openly conveying information i.e., being transparent (Garten, 2002).

Perception of Fairness

Perceived fairness is related to equity theory (Adams, 1965). Adams (1965), believed to be the founder of equity theory, postulated that the degree to which an exchange is equitable is determined by comparing the investment-outcome ratio of one person to the investment-outcome ratio of another person. People expect their investment-outcome ratio to be the same as the investment-outcome ratio of a referent other (Taris, Kalimo, & Schaufeli, 2002). If employee-A perceives his investment-outcome ratio is lower than employee-B's investment-outcome ratio, the fairness perception and morale of employee-A might be negatively affected. Fairness perceptions are closely related to organizational justice (Lambert et al., 2005) and the three dimensions of organizational justice i.e., procedural justice, distributive justice, and interactional justice (Peelle, 2007).

Procedural Justice

Early efforts to understand procedural justice focused on stipulating specific standards of process fairness. Germinal research on process fairness specified six criteria of fair procedures: consistency, bias suppression, accuracy, correctability, representativeness, and ethicality (Blader & Tyler, 2003). Subsequent research resulted in changes to the fair procedures criteria and added status recognition, trust in the benevolence of authorities, and neutrality (Blader & Tyler, 2003).

Procedural justice originated in a traditional model called organizational justice. Past organizational justice discussions examined perceptions of procedural justice and how those perceptions affect valued organizational outcomes (Gilliland, 1993). Procedural justice perceptions relative to organizational hiring and promoting practices are influenced by procedural rules, according to Gilliland (1993). Employee perceptions concerning the satisfaction or violation of procedural rules contribute to the

overall perception of the fairness of employee-selection (e.g.,
hiring and promoting) systems (Gilliland, 1993).

Procedural rules relative to an employee-selection system
might include considerations for job relatedness and feedback. Job
relatedness refers to the extent to which selection criteria is
relevant to the job, while feedback concerns the timely and
informative justifications given for finalized decisions. Two-way
communication during interviews (i.e., the opportunity offered to
ask questions and give input during interviews) has been found to
influence perceptions of procedural fairness (Gilliland, 1993).

Procedural justice, the perceived fairness of processes used
to determine outcomes, has been found to be influential in
promoting job satisfaction in the social service industry (Lambert
et al., 2005). To foster procedural justice, leaders should provide
employees an opportunity to express concerns about decisions and
the procedures for making decision (Lambert et al., 2005). Leaders
should share clear guidelines about how procedures should be

conducted and should conduct procedures honestly and with consideration for the views and rights of followers.

Distributive Justice

Distributive justice concerns dispensing rewards fairly (Adams, 1965; Lemons & Jones, 2001). Early research on distributive justice yielded the premise that people were more mindful of the fairness of outcomes (e.g., rewards) than of the level of outcomes (Colquitt et al., 2001). Unjust distributions by leadership can lead to perceptions of unfairness in some employees. In some cases, employees might accept an unequal distribution as fair and not feel unjustly treated. Adams offered, "If being a male sex is perceived as a higher investment than being of the female sex, a woman operator earning less than a man doing the same work will not feel unjustly treated" (Adams, 1965, p. 273).

Interactional Justice

Interactional justice concerns the quality of the interpersonal treatment (e.g., politeness and respect) an employee receives from leadership when procedures are applied (Colquitt et al., 2001). Information exchange is another aspect of interactional justice. The interactional justice of exchanging information relates to explaining why procedures were carried out and why rewards were distributed in a certain way (Colquitt et al., 2001). Perceptions of interactional justice have been found to play a role in employees' work attitudes and behavior. Employee satisfaction is enhanced when interactional justice is perceived (Colquitt et al., 2001).

Employee Morale

Research, concerning employee morale, conducted in the 1950s focused on employee needs which were believed to prompt drive and influence employees to strive to meet ends that satisfied those needs (Baehr & Renck, 1958). Morale, according to Baehr and Renck (1958), is structured by basic factors that include organization and management, material rewards, and job satisfaction. The organization and management factor deals with the employee's relationship with management and how the organization conveys information. The material rewards factor deals with the benefits (e.g., salary) employees receive for working, while job satisfaction concerns the intrinsic gratification associated with completing tasks that are believed to be worthwhile (Baehr & Renck, 1958). Employee morale, with respect to finalizing organizational procedures such as hiring and promoting decisions, might benefit from organizations openly sharing information and fostering employee perceptions of fairness.

Sashkin and Williams (1990) found, contrary to the results of research that support the notion that perceptions of fairness foster morale and productivity, that morale and productivity might not be nurtured in an environment where an employee believes leadership's actions are fair. An employee might agree that he is being treated equally fair when compared to another employee but be unhappy with his job. An employee could feel betrayed by a decision made by leadership that affects all employees equally. An organization might be forced to relegate all employees to part-time status. Although the decision affects employees equally, the decision may lead to disappointment in some employees and lower morale.

Baehr and Renck (1958) identified two factors that influence employee morale. The factors concern the employee's relationship with the formal organization and the employee's relationship with the supervisor. Employee sentiments of the organization and leadership's effectiveness and concern for the welfare of the employee regulate morale. Employees consider whether or not organizational leaders are friendly toward employees, live up to their promises, and solicit ideas from employees (Baehr & Renck, 1958). Effective leaders who show concern for their employees nurture morale and feelings of motivation within employees (Billikopf, n.d.).

Blanchard and Hersey's situational leadership theory suggests that leaders should employ flexible styles of leadership and be able to change their style when necessary to accommodate different situations in an effort to give directions and support to their followers (Hersey, Blanchard, & Johnson, 1996). Leaders who recognize what employees value and honestly seek to fulfill employee needs in exchange for favorable performance foster positive behaviors in employees (Lindner, 1998). Vroom's expectancy theory states that individuals can be motivated if favorable performance will result in a desirable reward and that individuals will pursue the level of performance that will generate the greatest reward (Hersey et al., 1996).

Transparency

Transparent is defined as (a) Letting light pass without distortion and (b) Free from pretense or deceit (Banff Executive Leadership Inc., 2003). The later definition is more relevant to the current research study. The concept of organizational transparency grew out of the need to stem corrupt practices.

Corruption was identified as an important problem that needed a solution (Transparency International, 2007b). Many international aid organizations, providing economic relief to countries, needed to ensure the financial aid they were providing was being used appropriately. The aid organizations requested open (i.e., transparent) accounting from all participating countries (Banff Executive Leadership Inc., 2003). Transparent practices can be applied within organizations to foster perceptions that procedures are being used appropriately.

Positive employee perceptions concerning hiring and promoting practices can be nurtured in organizations that foster transparency using committees to finalize selection decisions. Chan (2005) discussed the use of search committees to ensure faculty involvement in selections at a university. Chan (2005) wrote, "Search committees evolved out of a process to move away from more hierarchical managerial models of hiring" (p. 148). Older hierarchical hiring models fostered environments where the reproduction of the status-quo was the objective, i.e., hiring proved to be a "replication process for the dominant culture" (Chan, 2005, p. 159). Hiring models that focus on maintaining the status-quo are often viewed as unfair, may influence employee decisions to pursue discrimination cases (Gilliland, 1993), and may increase Federal legislation prohibiting unfair hiring practices.

Federal Equal Employment Opportunity (EEO) laws prohibiting job discrimination include Title VII of the Civil Rights Act of 1964, the Equal Pay Act of 1963, the Age Discrimination in Employment Act of 1967, and the Civil Rights Act of 1991 ("Federal Equal Employment Opportunity (EEO) Laws Prohibiting Job Discrimination", 2002). The Civil Service Reform Act (CSRA) of 1978, another law designed to foster fair employee selection practices, contains several prohibited personnel practices. The CSRA "prohibits any employee who has authority to take certain personnel actions from discriminating for or against employees or applicants for employment on the bases of race, color, national origin, religion, sex, age or disability" ("Federal

Equal Employment Opportunity (EEO) Laws Prohibiting Job

Discrimination", 2002. para. 2).

Organizations can guard against discrimination charges

relative to hiring and promoting processes by practicing

transparency. Garten (2002) described transparency as not being

opaque (i.e., being easily understood, clear, and candid).

Organizations can benefit from employing transparency while

developing and sharing policies and procedures (Garten, 2002).

Transparency reveals the internal controls used to govern the

organization and ensures all parties have access to organizational

information ("Baldrige glossary", 2007, Lainhart, 2000.)

The Aspen Institute conducts an annual conference in

Colorado to provide a venue for discussing and acting on critical

leadership issues such as transparency. Journalists, attending the

eighth annual Aspen Institute Conference on Journalism and

Society, agreed to increase transparency and open communication

throughout journalism in an effort to stem the growing distrust in

the media (Ziomek, 2005). The results of the 2004 presidential election initiated a barrage of media criticism relative to episodes of incomplete and inaccurate reporting. The subsequent decline in newspaper circulation led to apologies from leaders at major newspaper and promises to increase transparency by revealing sources or explaining why a source must remain in confidence, holding community forums, and publishing the email addresses of the reporters in an effort to regain the public's trust (Ziomek, 2005).

Current Findings

Modern researchers examine organizational concepts such as procedural justice and employee morale to realize relationships and behaviors that might benefit organizations. Some contemporary scholars continue to profess the importance of employing transparency to positively affect employee morale (Limas, 2005) and that using fair processes for hiring and promoting employees (e.g., using selection criteria to choose the most qualified applicant) will foster perceptions of fairness (Bagdadli et al., 2006). Modern researchers champion the beneficial use of practices that were maligned by historical researchers (Abdalla, Maghrabi, & Raggad, 1995) such as nepotism. Nepotism, derived from the Latin word for nephew (Johnson, 2003), is described as showing favoritism toward friends and relatives without giving an honest and objective evaluation of their abilities. The differing views of historical and modern researchers relative to employee selection practices and how those practices affect organizations establishes the need for further research in the area.

Theoretical Framework

The current research study focused on a three-part theoretical framework. The theoretical framework included procedural justice, employee morale, and transparency. Showing concern for the aforementioned three-part theoretical framework may enable organizational leaders to foster positive perceptions of organizational hiring and promoting processes and positively affect employee morale. The results of recent research, designed to yield factors relative to enhancing employee perceptions of fairness regarding promotion practices, supported the notion that using competency-oriented promotion criteria, providing well-defined promotion paths, and sharing information about promotion processes nurtures positive employee perceptions (Bagdadli et al., 2006).

Gilliland's research supported the belief that hiring and promoting procedures are perceived to be fairer when affected individuals have an opportunity to influence the process (Gilliland, 1993). Allowing employees to take part in organizational hiring and promoting processes may increase fairness perceptions and influence healthy organizational morale. The CEO of one of the world's largest staffing firms suggested that organizational leaders should ask their staff "for their thoughts on where the hiring need is greatest. They are on the front lines and often may notice gaps that aren't apparent. . ." (Messmer, 2005, p. 12).

Perception of Fairness

Increasing fairness perceptions is important for maintaining employee satisfaction. Arvey and Sackett (1993) hypothesized a relationship between perceived unfairness in an organizational system and higher rates of employee turnover. The perception of unfairness in hiring and promoting processes could reside in feelings of being the best qualified candidate but being overlooked for a position. Unfairness, contrastingly, could be perceived by employees who feel they were chosen for a position that should have been granted to a more qualified applicant.

Some scholars believe employee selections resulting from the use of criteria that is perceived as fair positively influences employee attitudes and leads to increases in organizational commitment (Bagdadli et al., 2006), while other scholars have argued that using selection criteria to finalized employee selections may not be a valuable organizational practice. The selecting official should, according to Miceli (n.d.), consider personal views as a determinant of which applicant should be selected. If a leader feels the most qualified applicant will not benefit the organization, the leader can justifiably select a lower qualified applicant. The concept of basing employee-selection decisions on personal views rather than strictly adhering to selection criteria is called prospect theory.

Prospect theory is described as establishing "a reference point which includes expectancies about the candidate's performance and how much the candidate can benefit the decision maker" (Miceli, n.d.). Prospect theory states that the selector can reject a candidate unless a clear demonstration of significant gain can be shown to result from hiring the candidate. Prospect theory identifies selectors as risk-averse under gain conditions and risk-prone under conditions where a gain is not evident (Miceli, n.d.). Relying on prospect theory may negatively affect employee perceptions of fairness. If employees perceive unfairness in the hiring and promoting processes, those employees might leave the company and seek fairer organizations (Sashkin & Williams, 1990).

Employees may develop an unfair perception of the organization if friends and relatives of leaders are selected to fill employment positions. Employees who feel they were more qualified than the selected relative or friend of an organizational leader often complain that nepotism was used in the decision making process. Some scholars believe selecting relatives for employment positions tends to foster a positive family-oriented environment and boosts morale for employees (Abdalla et al., 1995). Abdalla et al. (1995) listed several benefits of hiring family members such as working in a warm family-type environment, improved communications, and acceptance of a family-led organization by customers and the community. Benefits may be realized from selecting relatives, but employees might believe that promotions are given unjustifiably to relatives if proper selection procedures are not followed.

Procedural Justice

Recent procedural justice research supports the argument that procedural justice concerns extend beyond attention to how decisions are made (Blader & Tyler, 2003). Historical researchers stressed the importance of employing procedural rules when executing organizational processes. Modern researchers are beginning to increase focus on the quality of the treatment people experience as a party to organizational interactions relative to procedural justice considerations. When people evaluate the fairness of procedures they consider the way decisions are made and the type of treatment they experience as individuals (Blader & Tyler, 2003).

Employee Morale

Research reinforces the belief that employee morale can be negatively influenced by the application of unclear, unfair, and ambiguous organizational processes (Limas, 2005). Employee morale is negatively affected when leaders use processes to exact unnecessary control over their employees. Using employee performance appraisals as a form of control inhibits employee creativity, morale, and productivity (Limas, 2005). Morale can be contagious. Research results published in the last five years support the notion that employees who believe they have been victimized or discriminated against may not be productive to their organizations and can influence their coworkers to adopt feelings of lessening morale (Richardson, 2004).

Transparency

Transparency is perceived by team members to be instrumental in minimizing conflict. Transparency became a prominent objective of intra-governmental institutions for sharing leadership selection information. Transparency should be employed for the benefit of outside stakeholders and internal team members (Kahler, 2001). Kahler (2001) reported that leadership selections were virtually invisible to the public prior to 1990, but later criticisms led to the urgency to increase transparency.

Transparency is often prevented due to concerns of secrecy. The U.S. Department Of Defense (DOD) forgoes the practice of transparency in situations related to national security. Although the DOD has augmented business practices as a result of past scandals, concerns continue to exist relative to a lack of transparency (Transparency International, 2007a).

Walker and Pagano (2005) mentioned several requirements for building transparent leadership. Transparent leaders should be overwhelmingly honest (with respect and concern for others), gather intelligence (e.g., ask for feedback about their own performance and discover how others perceive them), deliver bad news well, and confess mistakes (Walker & Pagano, 2005). Somiah (2006) wrote, "Entrenching transparency in decision-making processes is a pre-emptive and offensive tactic. Documenting decision criteria and the processes followed in reaching decisions are measure(s) of transparency" (p. 141).

Explaining processes to employees fosters perceptions of fairness. An employee may not agree with an organizational decision, but find the decision to be acceptable due to a perception of fairness. Bies and Shapiro (1987) explained that people who receive negative outcomes tend to accept those results as fair when a reasonable explanation is offered. Leaders who offer fair and reasonable explanations promote transparency.

Rawlings (2008), after conducting research relative to organizational transparency, reported that organizations benefit from sharing information that allow stakeholders (e.g., employees) to make accurate decisions. Organizations should develop a trusting relationship with stakeholders and share information that is substantial and useful to the stakeholders (Rawlings, 2008). Trust should accompany transparency to foster open communications.

Literature Gaps

The current research study addresses literature gaps. Although research has suggested that perceived procedural justice by employees may lead to positive organizational outcomes and that perceptions of justice related to organizational hiring and promoting practices are influenced by procedural rules (Gilliland, 1993; Bagdadli et al., 2006), few studies have directly examined the relationship between employee perceptions concerning organizational hiring and promoting practices and employee morale. No studies have examined relationships between employee perceptions concerning organizational hiring and promoting practices and employee morale at a U.S. Army organization.

Research supports the notion that organizations benefit from being transparent about their policies and procedures (Garten, 2002) and that employing transparency positively affects employee morale (Limas, 2005). No studies have directly examined the relationship between organizational transparency and employee perceptions of fairness concerning hiring and promoting practices at a U.S. Army organization. One goal of the current research study was to fill the aforementioned gaps in the literature using a quantitative correlational research method that ensures anonymity to foster honest responses to the survey items.

Research Variables

This quantitative study included two types of variables (i.e., independent and dependent variables). The dependent variables of the current research study are employee perceptions of fairness concerning leaders' use of hiring and promoting practices and employee morale and the independent variable is organizational transparency. An independent variable can be manipulated to determine the extent of its relationship to a dependent variable. Dependent variables are sensitive to the changes of independent variables (Salkind, 2003).

The current research study examined the relationship between organizational transparency and employee perceptions of fairness. The study also examined employee morale as related to the perceptions employees hold concerning the fairness of leaders' use of hiring and promoting practices. Previous research has been conducted concerning employee perceptions of fairness and concerning a correlation between employee uncertainty and a decline in morale. The current research study added to previous research by revealing the degree to which employee perceptions of fairness relate to employee morale at an Army organization in Florida.

Conclusion

The literature review supported the notion that a relationship exists between organizational transparency and employee perceptions of fairness. The literature supported the postulation that a relationship exists between employee perceptions of fairness concerning leaders' use of hiring and promoting practices and employee morale. Employee perceptions of fairness relative to organizational hiring and promoting practices are influenced by procedural rules and employee needs (e.g., the need to balance the relationship between investments and outcomes). Needs, the focus of early research concerning morale, are believed to prompt drive and influence employees to strive to meet desired ends.

Past research reinforced the belief that employee morale can be negatively influenced by the lack of organizational transparency. Employee morale is negatively affected when leaders use processes to unnecessarily control their employees. Morale can be contagious and could benefit organizations if nurtured. Nurturing employee morale may lead to improved perceptions of fairness within organizations.

The literature review was helpful in buttressing the need to examine the degree to which employee perceptions of fairness relate to employee morale. The literature review facilitated examining the degree to which the fairness perceptions of the civilians employed at the Army organization in Florida correlate with the level of morale held by the employees. The literature review exposed gaps in research that justifies the need to examine the degree to which employees believe increased transparency in an organization's hiring and promoting practices improve perceptions of fairness.

Chapter Summary

Literature addressing the research questions was introduced in chapter two. The literature reviewed assisted in the formulation of the research hypotheses that a positive correlation exists between employee perceptions of fairness concerning leaders' use of fair hiring and promoting practices and employee morale and that increased transparency in organizational hiring and promoting processes greatly fosters employee perceptions of fairness. The methodology used in the current research study to confirm the reviewed literature and to support or reject the hypotheses will be detailed in chapter three. Chapter three will reveal the research design and examine the study participants.

CHAPTER 3: METHOD

The purpose of this study was to examine the relationship between employee perceptions of fairness, relative to organizational hiring and promoting practices, and employee morale. The effects of transparency on employee perceptions of fairness were also examined in the current research study. Historical research suggested that positive employee perceptions of leaders' decision making practices are nurtured when the processes that guide the finalization of decisions are transparent. Research supports the notion that decisions perceived as being made unfairly can result in employee dissatisfaction.

Chapter one addressed the notion that employee perceptions of fairness can influence morale and organizational productivity. Chapter one introduced the hypotheses and research questions of the current research study and described the methodology and theoretical body of knowledge that was used to answer the research questions. Chapter two introduced the body of knowledge that addressed the research questions and discussed the theoretical framework that guided this research. Chapter three will describe the methodology employed to answer the research questions and support or reject the hypotheses.

Research Design

The current research study used a quantitative research methodology and a Likert-type survey instrument to test the study hypotheses and answer the research questions with a .05 level of significance. Quantitative research methodologies are effectively used for problem-solving research concerning measurable variables (Creswell, 2005). The variables of the current research study (i.e., organizational transparency, employee morale, and employee perceptions of fairness concerning leaders' use of hiring and promoting practices) were measurable and were measured using a survey instrument. The survey instrument used within this research methodology facilitated the collection of nominal data representing the perceptions of a sample of employees at an Army organization. The research methodology process is outlined in the research methodology map (see Appendix A).

Appropriateness of Design

The research design used for this correlational study included a quantitative research methodology and a Likert-type survey instrument. Quantitative methodologies are appropriate for studies with known research variables (Creswell, 2005). Because the research variables for the current research study were known and the study examined relationships to reveal correlations, a quantitative approach was employed to test the research hypotheses and answer the research questions. Creswell noted that quantitative research studies can be used to describe correlations (2002).

A quantitative methodology was chosen for the current research study, rather than a qualitative one, because qualitative research methods are more effectively used when the research variables are unknown (Creswell, 2005). The objectivity offered by a quantitative methodology was beneficial for the current research study. Using a quantitative research methodology fosters the measurement of objective facts, while focusing on variables and reliability, and promotes researcher detachment (Neuman, 2003). The use of a Likert-type survey instrument as a part of this quantitative design facilitated an ordinal-level measure of each respondent's perspective relative to the hiring and promoting practices at the respondent's organization. Quantitative research allows the researcher to use a survey instrument designed to collect numeric data from respondents that can be statistically analyzed in an unbiased and objective manner (Creswell, 2005) and is conducive to objectively collecting, quantifying, and summarizing information to explain phenomena (Golafshani, 2003).

Research Questions

Maze (2007) mentioned unfair hiring practices, concerning military veterans, used by Federal managers. The practices included canceling job postings to avoid hiring veterans who were the top candidates and issuing complicated job listings that allowed multiple competitions for a single job (Maze, 2007). One veteran complained that it took 15 years to receive a promotion that should have been awarded years earlier. Promotions, according to the veteran, were continuously given to people with less experience (Maze, 2007).

A survey released in January, 2007 reported that only 57% of the Homeland Security employees were satisfied with their jobs. Only 67.5%, across the Federal Government, gave satisfactory reports concerning their jobs (Losey, 2007). Many Federal agencies are viewed as places where performers are rewarded for hard work, but the employees of some divisions of those agencies do not respect their leaders and report morale issues (Losey, 2007).

A Canadian news article included a report that managers in the Federal public service hire spouses, siblings and cousins rather than comply with rules to ensure all prospective applicants have a fair chance at Government jobs ("Public servants bend hiring rules, report finds", 2004). Similar perspectives have been reported in the U.S. by Federal employees (Losey, 2007). The Canadian article included a report on the findings of an investigative group that found numerous examples in which managers avoided open competitions to hire employees. A common theme in the review was that transactions seemed to have been structured from the outset with the goal of appointing a particular person ("Public servants bend hiring rules, report finds", 2004). Reflecting on these research findings, the following research questions were posed in the current research study:

Research question 1 (R1): What is the relationship between employee morale and the fairness perceptions of employees concerning organizational hiring and promoting practices?

Research question 2 (R2): What is the relationship between increased transparency in an organization's hiring and promoting practices and employee fairness perceptions?

The premise of this research study was that employee dissatisfaction, relative to perceptions of fairness and organizational transparency, affects employee morale within the organization. The current research study focused on the effects negative perceptions of organizational hiring and promoting practices have on employee morale. One objective of the research design chosen for the current research study was to employ an anonymous, but reliable, means of gathering employee perception data.

Population

The population for the current research study was composed of 592 Army civilians employed at an organization located in Florida. The civilian employees were part of a trained workforce where over 75% of the employees were certified in their employment specialty. The civilian employees constituting this population were required to complete 40 hours of annual training designed to prepare employees for future promotions. Many of the employees applied for upward mobility positions (i.e., promotions) within the organization.

Military personnel were not considered for the current research study (i.e., were not included in the target population). Military personnel reach higher ranks of the military hierarchy through excessive obedience (Dixon, 1976). Soldiers tend to refrain from sharing negative information concerning their superiors. Speaking negatively about superiors or the actions of superiors can result in demotions or other forms of punishment that block advancements.

One soldier shared negative information concerning his superiors' actions relative to prisoner abuse at the Abu Ghraib detention facility in Iraq. After testifying, the soldier reported being "stripped of his security clearance and blacklisted, making him ineligible for promotion, bonuses and military honors" ("US Congress subpoenas documents on Abu Ghraib abuse", 2006). Because of soldiers' fears of being severely punished for disobeying orders and sharing negative information concerning their perspectives of leaders' actions, military employee perspectives at the Army organization were not included in the current research study.

Informed Consent

Informed consent is an important ethical consideration in research involving human subjects (i.e., respondents). Obtaining the consent of respondents before allowing those respondents to take part in any research ensures agreement by all parties to voluntarily participate in a study. Respondents have the right to make decisions about what will and will not be done to them, while participating in a research study, and about what personal information they will share. Respondents should be given sufficient information to make a decision about their willingness to accept potential risks.

The current research study showed consideration for informed consent (see Appendix B). The cover letter that accompanied the survey instrument informed potential respondents that participation was voluntary. Assurances of confidentiality and measures taken to reduce risks associated with participation in the study were included in the survey instrument cover letter. The current research study presented minimal risk of harm to the respondents and the research did not involve procedures requiring consent outside the context of participation in the study.

Sampling Frame

The sample for the current research study was randomly chosen from the population of 592 civilian employees of an Army installation located in Florida. The sample was statistically significant and likely represented the normal perspectives of the population concerning the fairness of organizational hiring and promoting practices. The sample facilitated statistically significant and generalizable results. A sample that represents at least 50% of a population of 500 or more is accepted as representative of the population (Gay, 1996).

The random sample was chosen using 592 strips of paper with the name of a civilian employee on each piece of paper. The strips of paper containing the name of a civilian employee were deposited in a covered container and the container was shaken to cause the 592 pieces of paper to randomly reposition. Once the 592 pieces of paper settled into a final location, the container was

uncovered and a majority (i.e., at least 297) of the strips of paper

were blindly removed from the container. The names on the strips

of paper represented the employees randomly chosen to be

solicited as survey participants.

Confidentiality

Confidentiality refers to agreements made with survey

respondents concerning the protection of any provided

information. The current research study included multiple

safeguards to ensure the respondents' anonymity and

confidentiality were maintained. The research respondents were

required to give consent before responding to any survey items.

The respondents were required to respond to the survey items

anonymously, via a web-based survey instrument, to ensure their

decision to share their perceptions did not become a source of

potential harm. The consent and the survey data were collected

electronically and no physical forms were maintained.

Instrumentation

The sample data was gathered using a survey instrument adapted from a validated instrument used in a study that focused on the role of procedural justice in promotion decisions (Bagdadli et al., 2006). The resulting survey instrument for the current research study included a five-point Likert-type scale with 10 items seeking employee perception data. The survey items included: "Leaders at my organization use fair procedures in promotion decisions"; "Current hiring practices adversely affect my morale on the job"; "The best candidate for the job is usually hired, regardless of gender, race, or age"; and "If hiring and promoting processes were more openly shared with employees, my perception of fairness would improve." The survey instrument was validated via a pilot study (see Appendix D).

Pilot Study

A pilot study was used to validate the survey instrument. The pilot study entailed administering the survey instrument to 10 respondents selected from the population at the Army organization in Florida. The pilot study respondents were not asked to participate in the final study. Conducting a pilot study, according to Sproull (1995), offers the opportunity to assess the various research procedures prior to conducting the major study so that any necessary modifications could be made.

The pilot study respondents were asked to assess the content of the instrument for clarity, relevance, and specificity. The respondent's assessments were used to determine if the survey instrument was understandable and measured what it was designed to measure. Church and Waclawski (1998) informed that piloting a survey instrument should include an assessment that asks if the instructions and items are easy to read and understand, are the items meaningful to the participant, and are the items sufficiently detailed, or too general.

The results of the pilot study led to a need to delete item number seven. Item seven stated, "Performance appraisal ratings are considered fairly during promotion selections." A pilot survey respondent informed that the respondents would not know if the performance appraisal ratings were considered fairly during promotion selections and would not be able to provide a useful response to the survey item.

Item number six, revised as a result of the pilot study, stated, "The best candidate for the job is usually hired, regardless of gender, race, or age." A pilot survey respondent informed that the respondents would not know who interviewed for the job and would not know whether or not the best candidate was usually hired. Item six was changed to state, "It is my perception that the best candidate for the job is usually hired, regardless of gender, race, or age." The change to item six precipitated the usefulness in revising items one, two, three, four, five, and nine by adding the preface "It is my perception that. . ." to stress to the respondent that perceptions are acceptable for responding to the survey items. Following the pilot study and resulting changes to the survey instrument, a web-based version of the final survey instrument was created.

Instrument Reliability

Instrument reliability was examined using Cronbach's alpha computations. Cronbach's alpha computations measure the internal consistency of an instrument (Cronbach, 1951) and assess the extent that the items in a scale correlate with each other. Cronbach's alpha coefficients range from 0 to 1 and are accepted as good and as evidence of a reliable instrument when greater than .70 (Nunnally, 1978).

Reliability for the instrument used in this research was very strong and positive. The alpha coefficients for the survey items that focused on fairness, morale, and transparency perceptions were .87, .87, and .83 respectively (see Table 1). The computed Cronbach's alpha coefficients signify that the instrument used for the current research study was reliable and internally consistent.

Table 1

Instrument Reliability

Instrument focus area	Survey items	Number of items	Number of cases	Reliability coefficient (alpha)[a]	Confidence interval range
Fairness	1, 2, 4, and 6	4	64	.8687	.8069 - .9143
Morale	7 and 8	2	64	.8699	.7859 - .9210
Transparency	3, 5, and 9	3	64	.8341	.7488 - .8939

[a]All reliability coefficients have a 95.0% confidence level.

Data Collection

The respondent data was collected using a piloted and validated Likert-type five-point survey instrument designed to garner employee perception data. The survey instrument and accompanying cover letter were web-based and delivered to the personal email address of each prospective respondent (see Appendix D). The email message used to request study participation informed each recipient that participation in the survey was voluntary and required the respondent to select a hyperlink embedded in the message to confirm that the respondent was at least 18 years old and was voluntarily taking part in the survey. After selecting the hyperlink, the respondent was redirected to the electronic survey.

The survey remained open for four weeks. A reminder message was sent to the prospective respondents two weeks after the initial message was sent. The majority of the 64 respondents completed the survey during the first week. Only two surveys were completed during week three and no surveys were completed during week four. Permitting the survey to remain open for four weeks allowed employees who might have been out of the office (e.g., sick, vacationing, temporarily away on duty) an opportunity to participate in the study upon returning to the office.

The confidentiality of the respondents was protected by allowing the participants to respond to the survey items anonymously. Anonymity was allowed to ensure each respondent's decision to share his or her perceptions did not become a source of potential harm. The respondent data was collected electronically and no physical forms were maintained. The electronically collected data was stored in a password protected database and was deleted after being secured for three years.

Data Analysis

The information collected from the survey respondents was analyzed using standard statistical methods to summarize the research data. The data analysis was facilitated by using the Statistical Package for the Social Sciences (SPSS) analysis software program. The data was entered into the SPSS program to allow for statistical analysis and the creation of data charts and other presentation outputs. The data analysis process was designed to expose distribution central tendencies using the mean and mode. The mean, the most commonly used measure of central tendency (Salkind, 2003), shows the arithmetic average of responses to a survey item (Creswell, 2005) and the mode reveals the survey item response that was most frequently chosen (Salkind, 2003).

The Pearson product-moment correlation coefficient, the most often used correlation statistic (Glass & Hopkins, 1996), was computed to assess the level of association between the study variables. Several correlation coefficients were calculated to analyze the study data. The confidence level necessary for accepting each correlation as significant was 95.0% and the significance level was .05. Correlations with p-values of less than .05 support rejecting the null hypothesis with a 95.0% level of confidence. Correlations found between two variables denote that one variable can likely be used to predict the other variable, but does not prove causation (Sproull, 1995).

Reliability and Validity

Patton (2002) explained that validity and reliability are maximized in research designs that clearly explain the phenomenon being investigated while controlling potential biases or other factors that might distort the research results. Several factors have been identified as impacting validity and reliability. Factors that impact validity and reliability include the extent to which the sample is truly random, the size of the population, the desired margin of error of results, the desired confidence level of results, and the mitigation of non-response bias ("A White Paper on Research Sampling", 2005).

Summary

Chapter three described the methodology used in the current research study to answer the research questions. The methodology included validating a Likert-type survey instrument and issuing the survey to a population sample employed at an Army organization in Florida. Respondent data was electronically collected to promote anonymity. Respondent data was analyzed using the SPSS analysis software program.

The data analysis exposed central tendencies by computing means and modes. The data analysis included computing several Pearson product-moment correlation coefficients which resulted in strong correlations with p-values less than the expected .05. All correlations were significant and held p-values of .01 (i.e., supported rejecting each null hypothesis with a 99.0% level of confidence). Potential biases were minimized in the current research study by showing consideration for factors that may impact validity and reliability. Chapter four will present the results

of the data analysis and other considerations relative to the

research procedures outlined in Chapter three.

CHAPTER 4: PRESENTATION AND ANALYSIS OF DATA

The purpose of this quantitative correlational study was to examine the relationship between three variables: employee perceptions of fairness concerning organizational hiring and promoting practices, employee morale, and transparency. The research questions included in the current research study were;

R1: What is the relationship between employee morale and the fairness perceptions of employees concerning organizational hiring and promoting practices?

R2: What is the relationship between increased transparency in an organization's hiring and promoting practices and employee fairness perceptions?

The two sets of hypotheses advanced in the current research study were;

$H1_0$: A correlation does not exist between employee perceptions of fairness concerning leaders' use of fair hiring and promoting practices and employee morale,

H1: A correlation exists between employee perceptions of fairness concerning leaders' use of fair hiring and promoting practices and employee morale,

$H2_0$: Increased transparency in organizational hiring and promoting processes does not greatly foster employee perceptions of fairness,

H2: Increased transparency in organizational hiring and promoting processes greatly fosters employee perceptions of fairness.

Chapter one addressed the problem that employee perceptions of fairness, influenced by leaders' use of hiring and promoting practices and transparency, may correlate with organizational morale. Chapter one introduced research that examined the extent of the relationship between employee perceptions of fairness concerning organizational hiring and promoting practices and employee morale and research that examined the relationship between organizational transparency and employee perceptions of fairness. Chapter two provided a review of the literature; addressed the research questions and key issues relative to employee perceptions of leaders' hiring and promoting practices, employee morale, and organizational transparency; and introduced the hypotheses that a correlation exists between employee perceptions of fairness concerning leaders' use of fair hiring and promoting practices and employee morale and that increased transparency in organizational hiring and promoting processes greatly fosters employee perceptions of fairness. Chapter three described the methodology employed to answer the research questions and introduced the data collection and analysis procedures. Chapter four will present an analysis of the data collected (i.e., an analysis of the responses given by the subjects to the survey items included in the survey instrument).

Demographics

The population sample for the study was randomly chosen from the 592 civilian employees of the Army installation located in Florida. Sixty percent of the population was male and 40% was female. Table 2 displays population demographics information by gender.

Table 2

Demographics Summary-Gender

Demographics Summary-Gender		
Gender	N	Population percentage
Male	355	60
Female	237	40
Total	592	100

Five percent of the organization's population consists of active duty military and 95.0% consist of non-military personnel (i.e., civilians). Although the population at the Army installation consisted of 31 active duty military personnel, only the civilian population was sampled and solicited to take part in the current research study. Table 3 displays population demographics information by status (i.e., relative to whether or not the employee is civilian or active duty military).

96

Table 3

Demographics Summary-Status

Demographics Summary-Status		
Status	N	Population percentage
Active duty military	31	5
Civilians	592	95
Total	623	100

Final Study

The final version of the web-based survey instrument was offered to 300 civilians employed at an Army organization located in Florida. Sixty-four of the civilian employees responded to the web-based survey instrument. The survey instrument is located in an appendix to the current research study (see Appendix D). The responses to the survey items submitted by the 64 civilians employed at the Army organization based in Florida are located in

Table 8 as an appendix to the current research study (see Appendix E).

Data Analysis and Results

The research data for the current research study was collected using a web-based five-point Likert-type survey instrument with 10 items seeking employee perception data. The survey instrument contained measures ranging from *strongly disagree* to *neither agree or disagree* to *strongly agree.* Each measure corresponded to a number from one to five. Each *strongly disagree* selection represented a one, each *disagree* selection equated to a two, each *neither agree or disagree* selection corresponded to a three, each *agree* selection matched a four, and each *strongly agree* selection represented a five.

The respondent data was manually entered into the Statistical Package for the Social Sciences (SPSS) software analysis program data editor spreadsheet. Sixty-four rows were populated with the numbers ranging from one to five to represent each respondent's survey item response selection. The 10 spreadsheet columns represent the 10 survey items. A replica of the SPSS data editor spreadsheet with respondent data is included in Table 8 as an appendix to the current research study (see Appendix E). Combinations of data-filled columns were imported into the SPSS analysis and graphical presentation tools for the data analyses and results presentations included in this chapter.

The research data was analyzed, using the SPSS software program, to expose the mode and mean of the distributions and the variable correlations. The mode was computed for each survey item to ascertain the survey item response(s) that occurred most frequently and the mean was computed to show the arithmetic average of responses to each survey item. Variable correlations were exposed, using Pearson product-moment correlation coefficients, to examine the relationship between employee perceptions of fairness concerning leaders' use of hiring and promoting practices and employee morale. The correlation coefficients also examined the relationship between organizational transparency and employee perceptions of fairness concerning leaders' use of hiring and promoting practices.

Correlation coefficients reveal the strength of linear relationships between research variables (e.g., correlation coefficients can reveal how closely multiple survey response variances match). Correlation coefficient values range from -1.0 to +1.0 and are accepted as positive if the value is a positive number and negative if the value is a negative number. Strong positive correlations are deemed present when the correlation coefficient approaches a value of +1.0 and strong negative correlations are accepted as present when the correlation coefficient approaches a value of -1.0. A correlation coefficient of 0 indicates that no relationship between variables is present. Any strong correlations found denote that one variable is likely a predictor of the other variable, although the assumption that a change in one variable caused a change in the other variable cannot be assumed (Sproull, 2004).

Hypotheses Testing Results

Two sets of hypotheses consisting of null and alternate hypotheses were advanced in the current research study.

H1 null hypothesis stated a correlation does not exist between employee perceptions of fairness concerning leaders' use of fair hiring and promoting practices and employee morale.

H1 alternate hypothesis stated a correlation exists between employee perceptions of fairness concerning leaders' use of fair hiring and promoting practices and employee morale.

H2 null hypothesis stated increased transparency in organizational hiring and promoting processes does not greatly foster employee perceptions of fairness.

H2 alternate hypothesis stated increased transparency in organizational hiring and promoting processes greatly fosters employee perceptions of fairness.

hypothesis one. The H1 null hypothesis (A correlation does not exist between employee perceptions of fairness concerning leaders' use of fair hiring and promoting practices and employee morale) was tested by computing a correlation coefficient and analyzing the coefficient to ascertain the existence of any linear relationship between the variables.

The results of the data analysis are presented below in Figures 1 and 2. Several respondents chose the same responses for the survey items compared in Figures 1 and 2 which resulted in several of the data points (i.e., dots) displayed in Figures 1 and 2 being superimposed (i.e., representing multiple responses). The computed correlations were significant at the .05 level.

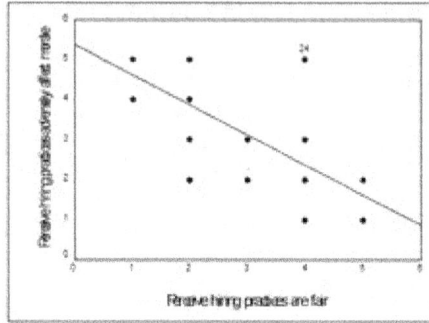

Figure 1. Correlational representation of
employee perceptions of leaders' use of fair
hiring practices and employee perceptions of
the adverse affect of hiring practices on
employee morale.

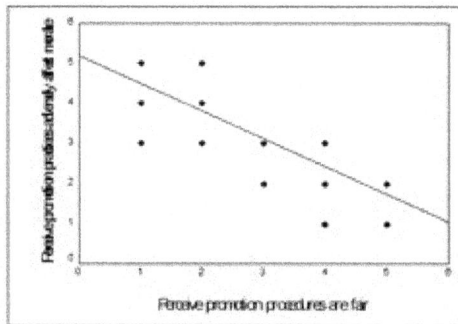

Figure 2. Correlational representation of
employee perceptions of leaders' use of fair
promotion procedures and employee

perceptions of the adverse affect of promotion

practices on employee morale.

A significance level of .05 denotes that the chance of rejecting the null hypothesis when the null hypothesis is true is .05 or 5.0%. The significance probability (i.e., p-value) for each correlation was .01. When the p-value of a statistical test is lower than the significance level, the null hypothesis can be rejected with a low probability of error (Sterne & Smith, 2001). The p-values for the correlations shown in Figures 1 and 2 were less than the significance levels and supported rejecting the null hypothesis.

The H1 null hypothesis was tested in two steps. In step one, respondent responses to the survey instrument item related to perceptions of fairness concerning leaders' use of fair hiring procedures were compared to responses given to the survey item concerning whether or not hiring practices adversely affect morale. The responses were imported into the SPSS analysis tool and a Pearson product-moment correlation coefficient was computed to show the extent of any relationship between perceptions of fairness concerning leaders' use of fair hiring procedures and whether or not hiring practices adversely affect morale.

Step two for testing the H1 null hypothesis entailed comparing respondent responses to the survey instrument item related to perceptions of fairness concerning leaders' use of fair promoting procedures with responses given to the survey item concerning whether or not promotion practices adversely affect morale. The responses were imported into the SPSS analysis tool and a Pearson product-moment correlation coefficient was computed to show the extent of any relationship between perceptions of fairness concerning leaders' use of fair promotion procedures and whether or not promotion practices adversely affect morale.

The data analysis results displayed in Figure 1 show a strong negative correlation between employee perceptions of leaders' use of fair hiring practices and employee perceptions of the adverse affect of the hiring practices on employee morale. The correlation coefficient corresponding to the data analysis results displayed in Figure 1 was -.72. A correlation is accepted as strong

when the correlation coefficient ranges from .70 to 1.0 or from -.70

to -1.0 (Dunn, 2000).

The correlation between employee perceptions of leaders'

use of fair hiring practices and employee perceptions of the

adverse affect of the hiring practices on employee morale was

significant at the .05 level. The confidence level to accept the

correlation as significant was 95.0%. The p-value for this

correlation was .01 (see Table 4). Correlations with p-values of .01

support rejecting the H1 null hypothesis with a 99.0% level of

confidence.

Table 4

Statistical Significance of Fair Hiring-Morale Affect Correlation

	Fair hiring practices perception	Adverse morale affect
Fair hiring practices perception		
Pearson correlation	1	-.723**
N	64	64

**$p < .01$.

Analysis of the respondents' responses (i.e., raw data)
revealed that 28% of the employees who selected strongly disagree
or disagree to survey item one which read "It is my perception that
leaders at my organization use fair procedures in hiring decisions"
also selected agree or strongly agree to survey item seven which
read "Current hiring practices adversely affect my morale on the
job." Thirty-one percent of the employees who selected agree or
strongly agree to survey item one also selected strongly disagree or
disagree to survey item seven. Figure 1 graphically displays the
linear relationship.

Figure 2 shows a strong negative correlation between
employee perceptions of leaders' use of fair promoting practices
and employee perceptions of the adverse affect of those promoting
practices on employee morale. The correlation coefficient
corresponding to the data analysis results displayed in Figure 2
was -.81. A correlation is accepted as strong when the correlation
coefficient ranges from .70 to 1.0 or from -.70 to -1.0 (Dunn,
2000).

The correlation between employee perceptions of leaders'
use of fair promoting practices and employee perceptions of the
adverse affect of the promoting practices on employee morale was
significant at the .05 level and the confidence level for accepting
the correlation as significant was 95.0%. The p-value for this
correlation was .01 (see Table 5). Correlations with p-values of .01
support rejecting the H1 null hypothesis with a 99.0% level of
confidence.

Table 5

Statistical Significance of Fair Promoting-Morale Affect
Correlation

	Fair promoting practices perception	Adverse morale affect
Fair promoting practices perception		
Pearson correlation	1	-.805**
N	64	64

**p < .01

Analysis of the respondents' responses (i.e., raw data)
revealed that 34% of the employees who selected strongly disagree
or disagree to survey item two which read "It is my perception that
leaders at my organization use fair procedures in promotion
decisions" also selected agree or strongly agree to survey item

eight which read "Current promotion practices adversely affect my morale on the job." Twenty-two percent of the employees who selected agree or strongly agree to survey item two also selected strongly disagree or disagree to survey item eight. Figure 2 graphically displays the linear relationship.

Figures 1 and 2 show correlations that are significant at the .05 level. A significance level of .05 denotes that the chance of rejecting the null hypothesis when the null hypothesis is true is .05 or 5.0%. The significance probabilities (i.e., p-values) for the correlations were .01 and supported rejecting the H1 null hypothesis with a low probability of error because the p-values were less than the significance level of .05.

hypothesis two. The H2 null hypothesis (Increased transparency in organizational hiring and promoting processes does not greatly foster employee perceptions of fairness) was tested via data analysis using the SPSS analysis tool. After computing the means of survey items one, two, three, four, five, and six to ascertain the average response chosen for each survey item, the modes for the six items were manually computed to reveal the answer most often chosen for each item. The results of the data analysis are shown in Table 9 (see Appendix F).

The raw data analysis results displayed in Table 9 of Appendix F show that the majority of the respondents selected neither agree or disagree in response to survey item 10 (i.e., the mode for survey item 10 was three). Survey item 10 stated, "If explanations of the hiring and promoting processes were more openly shared with employees, my perception of fairness would improve." While 50.0% of the respondents selected neither agree or disagree, 41.0% of the respondents agreed that increased transparency relative to hiring and promoting processes would improve their perception of fairness. The individual responses for survey item 10 are located in Table 8 (see Appendix E).

Research Questions

One objective of the current research study was to answer two research questions. The answers to the two research questions were needed to facilitate an examination of the relationships

between organizational transparency, employee perceptions, and

morale relative to leaders' use of hiring and promoting practices at

an Army organization in Florida. The research questions posed in

the current research study were R1: What is the relationship

between employee morale and the fairness perceptions of

employees concerning organizational hiring and promoting

practices and R2: What is the relationship between increased

transparency in the organization's hiring and promoting practices

and employee fairness perceptions?

research question one. The first research question (What is

the relationship between employee morale and the fairness

perceptions of employees concerning organizational hiring and

promoting practices?) was answered after analyzing the respondent

data. The results of the data analysis are presented in Figures 1 and

2. The analyses show that strong correlations exist between

employee perceptions of leaders' use of fair hiring and promotion

practices and employee perceptions of the adverse affect of the hiring and promotion practices on employee morale. The strong

correlations found were statistically significant at the .05 level and

supported rejecting the H1 null hypothesis with more than a 95.0%

level of confidence.

Analysis of the respondents' responses (i.e., raw data)

revealed that 27% of the respondents that disagreed with survey

item one, indicating they do not perceive leaders use fair hiring

procedures, also agreed with survey item eight, indicating that

current hiring practices adversely affect morale. Thirty-one percent

of the respondents that agreed with survey item one, indicating

they perceive leaders use fair hiring procedures, also disagreed

with survey item eight, indicating that current hiring practices do

not adversely affect morale. Thirty-three percent of the

respondents that disagreed with survey item two, indicating they

do not perceive leaders use fair promotion procedures, also agreed

with survey item eight, indicating that current promotion practices

adversely affect morale. Twenty-two percent of the respondents

that agreed with survey item two, indicating they perceive leaders use fair promotion procedures, also disagreed with survey item eight, indicating that current promotion practices do not adversely affect morale. The individual responses to the survey items that were used for these results are located in Table 8 as an appendix to the current research study (see Appendix E).

research question two. The second research question posed by the current research study (What is the relationship between increased transparency in the organization's hiring and promoting practices and employee fairness perceptions?) was answered after analyzing the respondent data in two steps. In step one, respondent responses to the survey instrument item related to perceptions of fairness concerning leaders' use of fair hiring procedures were compared to responses given to the survey item concerning whether or not hiring and promoting processes are openly shared with employees. The responses were imported into the SPSS analysis tool and a Pearson product-moment correlation coefficient was computed to show the extent of any relationship between perceptions of fairness concerning leaders' use of fair hiring procedures and whether or not hiring and promoting processes are openly shared with employees. The second step in the data analysis entailed comparing respondent responses to the survey instrument item related to perceptions of fairness concerning leaders' use of fair promoting procedures to responses given to the survey item concerning whether or not hiring and promoting processes are openly shared with employees. The responses were imported into the SPSS analysis tool for the analysis and a Pearson product-moment correlation coefficient was computed to show the extent of any relationship between perceptions of fairness concerning leaders' use of fair promoting procedures and whether or not hiring and promoting processes are openly shared with employees.

The results of the two-part respondent data analysis are displayed in Figures 3 and 4. The results revealed positive correlations that were statistically significant at the .05 level with p-values of .01. (see Table 6). The p-values for the correlations supported rejecting the H2 null hypothesis with a 99.0% level of confidence. Several respondents chose the same responses for the

survey items compared in Figures 3 and 4 which resulted in

several of the data points (i.e., dots) displayed in the figures being

superimposed (i.e., representing multiple responses).

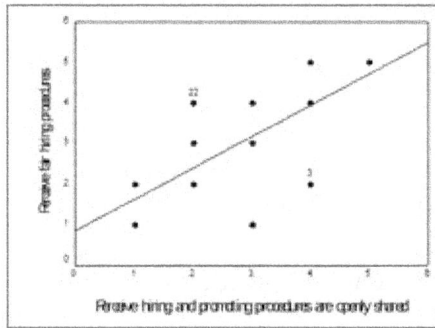

Figure 3. Correlational representation of

employee perceptions of leaders' use of fair

hiring procedures and employee perceptions

of whether or not hiring and promoting

procedures are openly shared.

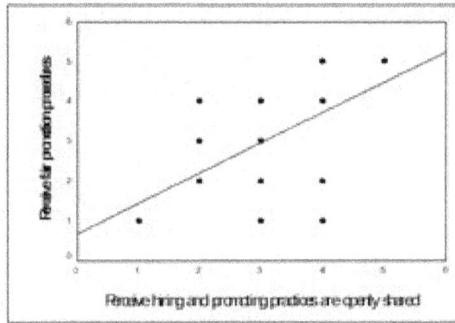

Figure 4. Correlational representation of

employee perceptions of leaders' use of fair

promotion procedures and employee

perceptions of whether or not hiring and

promoting procedures are openly shared.

The data analysis results displayed in Figure 3 show a

strong positive correlation between employee perceptions of

leaders' use of fair hiring procedures and employee perceptions of

whether or not hiring and promoting procedures are openly shared.

The correlation coefficient corresponding to the data analysis

results displayed in Figure 3 is .71. The correlation between

employee perceptions of leaders' use of fair hiring procedures and

116

employee perceptions of whether or not hiring and promoting

procedures are openly shared was significant at the .05 level.

Table 6

Statistical Significance of Fair Hiring-Shared Procedures Correlation

	Fair hiring procedures perception	Procedures openly shared
Fair hiring procedures perception		
Pearson correlation	1	.705**
N	64	64

**p < .01

Analysis of the respondents' responses (i.e., raw data)

revealed that 28% of the employees selected agree or strongly

agree to survey item nine which read "It is my perception that

hiring and promoting processes are openly shared with

employees." Those same respondents selected agree or strongly

agree to survey item one which read "It is my perception that

leaders at my organization use fair procedures in hiring decisions."

Twenty-eight percent of the employees who selected strongly

disagree or disagree to survey item nine also selected strongly

disagree or disagree to survey item one.

Two of the respondents selected strongly disagree or disagree to survey item nine which read "It is my perception that hiring and promoting processes are openly shared with employees", but selected agree or strongly agree to survey item one which read "It is my perception that leaders at my organization use fair procedures in hiring decisions." Four of the respondents selected agree or strongly agree to survey item nine, indicating their perception that hiring and promoting processes are openly shared at the Army organization, but selected strongly disagree or disagree to survey item one, indicating their perception that leaders at the Army organization in Florida use fair procedures in hiring decisions. The contrary responses provided by the six respondents did not significantly skew the results of the research data analysis. The responses, accepted as outlier research data, will be discussed in chapter five relative to why the responses given by the six respondents were contrary to the responses given by the remaining 58 respondents.

The data analysis results displayed in Figure 4 show a weak positive correlation between employee perceptions of leaders' use of fair promotion procedures and employee perceptions of whether or not hiring and promoting procedures are openly shared. The correlation coefficient corresponding to the data analysis results displayed in Figure 4 is .64. The correlation between employee perceptions of leaders' use of fair promoting procedures and employee perceptions of whether or not hiring and promoting procedures are openly shared was significant at the .05 level with p-values of .01. (see Table 7).

Table 7

Statistical Significance of Fair Promoting-Shared Procedures Correlation

	Fair promoting procedures perception	Procedures openly shared
Fair promoting procedures perception		
Pearson correlation	1	.645**
N	64	64

**p < .01*

Analysis of the respondents' responses (i.e., raw data) revealed that 25% of the employees selected agree or strongly agree to survey item number nine which read "It is my perception that hiring and promoting processes are openly shared with employees." The same respondents selected agree or strongly agree to survey item two which read "It is my perception that leaders at my organization use fair procedures in promotion decisions." Twenty-eight percent of the employees who selected strongly disagree or disagree to survey item nine also selected strongly disagree or disagree to survey item two.

Two of the 64 respondents, contrary to the correlation shown in Figure 4, selected strongly disagree or disagree to survey item nine and selected agree or strongly agree to survey item two. Six of the 64 respondents selected agree or strongly agree to survey item nine and selected strongly disagree or disagree to survey item two. Two of the six respondents who selected agree or strongly agree to survey item nine selected strongly disagree or disagree to survey item five which read "It is my perception that promotion opportunities are well communicated through formal channels to potential candidates." Discussions of the contrary responses will be included in chapter five of the current research study.

Outliers

The outlier data, represented in Figures 1 and 3 as dots beneath Arabic numbers, symbolize contrary responses (i.e., response data that falls outside of the central tendencies). Outliers can be rejected if they prove to be values resulting from measurement error that distort the analysis or outliers can be accepted as phenomena of interest (Redpath & Sheard, 2005). Although statistical measures can be sensitive to the presence of outliers, the outliers in Figures 1 and 3 do not appear to result from measurement error nor do they distort the findings of correlation between the variables compared to test the hypotheses and answer the research questions. The outlier data in Figures 1 and 3 were accepted.

Summary

Chapter four presented a data analysis of the responses to

the survey items included in the survey instrument. The SSPS software tool was used to analyze the research data. The analysis exposed distribution central tendencies and variable correlations to test the research hypotheses and answer the research questions.

The examinations of the hypotheses were facilitated by the data analysis results revealed in chapter four. The H1 null hypothesis stated a correlation does not exist between employee perceptions of fairness concerning leaders' use of fair hiring and promoting practices and employee morale and the H2 null hypothesis stated increased transparency in organizational hiring and promoting processes does not greatly foster employee perceptions of fairness. The H1 and H2 null hypotheses were rejected as a result of the data analyses of the current research study. The data analyses revealed strong correlations between employee perceptions of fairness concerning leaders' use of fair hiring and promoting practices and employee morale. The data analyses showed correlations between employee perceptions of leaders' use of fair hiring and promoting practices and employee perceptions of whether or not hiring and promoting procedures are openly shared (i.e., are transparent).

Chapter five will focus on fairness perceptions, morale, and transparency while offering interpretations of the results reported in chapter four. Actions will be proposed to assist in fostering fairness perceptions, morale, and transparency. Chapter five will include recommendations for further research.

CHAPTER 5: CONCLUSIONS AND

RECOMMENDATIONS

Chapter one addressed the problem that employee perceptions of fairness, influenced by leaders' use of hiring and promoting practices and transparency, may correlate with organizational morale and introduced research examining the extent of the relationships between organizational transparency, employee perceptions of fairness concerning organizational hiring and promoting practices, and employee morale. Chapter two provided a review of the literature; addressed the research questions and key issues relative to employee perceptions of leaders' hiring and promoting practices, employee morale, and organizational transparency; and introduced the hypotheses that a correlation exists between employee perceptions of fairness concerning leaders' use of fair hiring and promoting practices and employee morale and that increased transparency in organizational hiring and promoting processes greatly fosters employee perceptions of fairness. Chapter three described the methodology employed to answer the research questions and introduced the data collection and analysis procedures. Chapter four presented an analysis of the data collected exposing distribution central tendencies and variable correlations to test the research hypotheses and answer the research questions. Chapter five will focus on fairness perceptions, morale, and transparency, while offering interpretations of the results reported in chapter four, and suggest recommendations for further research.

The results of a survey released in 2007 supported the position that less than 60.0% of the employees at a Federal Government organization were satisfied with their jobs. Only 67.5%, across the Federal Government, gave satisfactory reports concerning their jobs (Losey, 2007). Scholars, seeking causes of less than optimal job-satisfaction reports and ways to improve organizational relationships, reported research findings supporting relationships between employee perceptions and employee attitudes (Bagdadli, Roberson, & Paoletti, 2006; Makawatsakul & Kleiner, 2003).

The purpose of this quantitative correlational research study was to examine the relationships between two dependent variables (i.e., the relationship between employee morale and employee perceptions of fairness concerning organizational hiring and promoting practices) and one independent variable (i.e., transparency). A web-based five-point Likert-type survey instrument was used to gather employee perception data relative to the dependent and independent variables. A sample of the 592 civilians employed at an Army organization in Florida was surveyed to gather perception data that was analyzed to test the hypotheses and answer the research questions of the current research study. The interpretations of the research data analysis will offer perspectives that might assist in resolving problems relative to unfavorable employee perceptions fostered by the lack of openly shared procedural information.

Findings and Interpretations

Two sets of hypotheses consisting of null and alternate hypotheses were advanced in the current research study. The H1 null hypothesis stated a correlation does not exist between employee perceptions of fairness concerning leaders' use of fair hiring and promoting practices and employee morale. The H1 alternate hypothesis stated a correlation exists between employee perceptions of fairness concerning leaders' use of fair hiring and promoting practices and employee morale. The H2 null hypothesis stated increased transparency in organizational hiring and promoting processes does not greatly foster employee perceptions of fairness. The H2 alternate hypothesis stated increased transparency in organizational hiring and promoting processes greatly fosters employee perceptions of fairness.

Two research questions were posed in the current research study. The first research question inquired about the relationship between employee morale and the fairness perceptions of employees concerning organizational hiring and promoting practices. The second research question inquired about the relationship between increased transparency in the organization's hiring and promoting practices and employee fairness perceptions.

The findings of the data analyses, relative to the hypotheses and research questions, are summarized below.

Hypothesis One

The H1 null hypothesis ($H1_0$) was tested by computing a correlation coefficient and analyzing the coefficient to ascertain the existence of any linear relationship between the variables. The results of the data analysis revealed a strong negative correlation between employee perceptions of fairness concerning leaders' use of fair hiring and promoting practices and employee morale. The correlation was significant at the .05 level and justified rejecting $H1_0$ which stated that a correlation does not exist between employee perceptions of fairness concerning leaders' use of fair hiring and promoting practices and employee morale. Correlations found to be significant at the .05 level offer support for the research hypothesis and support rejecting the null hypothesis with a 5.0% chance of making an erroneous rejection.

Analysis of the respondents' responses (i.e., raw data) revealed that 28% of the Army civilians who disagreed that they held the perception that leadership at their organization use fair hiring procedures agreed that current hiring practices adversely affect their morale on the job. Thirty-one percent of the employees who agreed that they held the perception that leadership at their organization use fair hiring procedures disagreed with the suggestion that current hiring practices adversely affect their morale on the job. These findings support the notion that when leadership fosters employee perception of the use of fair hiring procedures employee morale is positively affected.

Thirty-four percent of the Army civilians who disagreed that they held the perception that leadership at their organization use fair promoting procedures agreed that current promotion practices adversely affect their morale on the job. Twenty-two percent of the employees who agreed that they held the perception that leadership at their organization use fair promoting procedures disagreed with the suggestion that current promotion practices adversely affect their morale on the job. These findings support the notion that when leadership fosters employee perception of the use of fair promoting procedures employee morale is positively affected.

H1 null hypothesis data analysis findings support the findings of Messer and White (2006) that perceived fairness significantly affects organizational effectiveness and performance and the postulation made by Makawatsakul and Kleiner (2003) offering that employee perceptions of organizational leadership's use of unfair hiring and promoting processes may degrade employee morale. Degraded morale might have been the reason for the results of a survey released in 2007 that supported the position that less than 60.0% of the employees at a Federal Government organization were satisfied with their jobs. Only 67.5%, across the Federal Government, gave satisfactory reports concerning their jobs (Losey, 2007). Scholars, seeking causes of less than optimal job-satisfaction reports and ways to improve organizational relationships, reported research findings supporting relationships between employee perceptions and employee attitudes (Bagdadli, Roberson, & Paoletti, 2006; Makawatsakul & Kleiner, 2003).

This research supports germinal findings that perceived fairness is related to equity theory, i.e., employees' feelings that they are receiving equal treatment when compared to other employees (Adams, 1965). When an employee works as hard as a coworker, the employee expects to have an equal opportunity relative to receiving a promotion. Some employees experience feelings of reduced morale when promotions or hiring advantages are perceived as being given to others unfairly, e.g., nepotism-based promotions (Makawatsakul & Kleiner, 2003).

The results of the research support the notion that leaders who are perceived to be employing unfair hiring and promoting practices may be harming the organization by reducing employee morale. Although some scholars believe selecting relatives for employment positions tends to foster a positive family-oriented environment and boosts employee morale (Abdalla et al., 1995), employee and organizational morale may suffer if the selection is perceived as unfair by employees. Miceli (n.d.) suggested that basing hiring and promoting decisions on expectancies about the candidate's performance and how much the candidate can benefit the decision maker (i.e., prospect theory) can help an organization.

The results of the current research study do not support the postulation that using the concept of prospect theory to select employees can help an organization. The results of the current research study support the hypothesis that a positive correlation exists between employee perceptions of fairness concerning leaders' use of fair hiring and promoting practices and employee morale. The results support the belief that fair selection procedures (e.g., pre-established criteria-based selections) nurture morale.

Hypothesis Two

The H2 null hypothesis was tested by computing the means of survey items one, two, three, four, five, and six to ascertain the average response chosen for each survey item. The modes for the six items were computed to reveal the answer most often chosen for each item. The mean for each of the six items was three and the modes ranged from two to four (see Appendix F). The responses given to the survey item that suggested openly sharing hiring and promoting processes would improve fairness perceptions (i.e., survey item 10) revealed that 41.0% of the Federal employees surveyed agreed that increased transparency in hiring and promoting processes would improve their perception of fairness.

The mean for survey item 10 was three (neither agree or disagree). If the responses of neither agree or disagree to survey item 10 are accepted as irrelevant to testing the H2 null hypothesis, due to being non-committal, and those responses are disregarded, the mean for survey item number 10 would change from three to four (agree). The representative mean of four for survey item number 10 supports the hypothesis that increased transparency relative to hiring and promoting processes would improve perceptions of fairness.

The results of the data analysis relative to the H2 null hypothesis of the current research study support the findings of several scholars. The data analysis results concerning the H2 null hypothesis support the findings of Limas (2005) who postulated that employing transparency positively affects employee morale. Other scholars concluded that organizations could benefit from employing transparency while developing and sharing policies and procedures such as using selection criteria to choose the most qualified applicant (Bagdadli et al., 2006; Garten, 2002). The results of the current research study support rejecting the H2 null hypothesis and accepting the H2 alternate hypothesis that increased transparency in organizational hiring and promoting processes greatly fosters employee perceptions of fairness.

Research Question One

The results of the data analysis revealed strong negative correlations between employee perceptions of leaders' use of fair hiring and promotion practices and employee perceptions of the adverse affect of the hiring and promotion practices on employee morale. Twenty-seven percent of the respondents who indicated they do not perceive leaders use fair hiring procedures also indicated that current hiring practices adversely affect morale. Thirty-one percent indicated they perceive leaders use fair hiring procedures, but also indicated that current hiring practices do not adversely affect morale. Thirty-three percent indicated they do not perceive leaders use fair promotion procedures, but also indicated that current promotion practices adversely affect morale. Twenty-two percent of the respondents who indicated they perceive leaders use fair promotion procedures also indicated that current promotion practices do not adversely affect morale.

Research Question Two

The results of the data analysis showed that positive correlations exist between employee perceptions of leaders' use of fair hiring and promoting procedures and employee perceptions of whether or not hiring and promoting procedures are openly shared.

Twenty-eight percent of the employees who indicated that hiring and promoting processes are openly shared with employees also indicated that leaders at their organization use fair procedures in hiring decisions. Twenty-eight percent of the employees who indicated that hiring and promoting processes are not openly

shared with employees also indicated that leaders at their

organization do not use fair procedures in hiring decisions.

Twenty-five percent of the employees who indicated that hiring

and promoting processes are openly shared with employees also

indicated that leaders at their organization use fair procedures in

promotion decisions. Twenty-eight percent of the employees who

indicated that hiring and promoting processes are not openly

shared with employees also indicated that leaders at their

organization do not use fair procedures in promotion decisions.

Outliers

Several of the research subjects selected responses that

were contrary to the responses given by the other subjects. Some

respondents disagreed that hiring and promoting processes are

openly shared with employees, but agreed that leaders at their

organization use fair hiring and promoting procedures. Other

respondents agreed that hiring and promoting processes are openly

shared with employees, but disagreed that leaders at their

organization use fair hiring and promoting procedures. While the possibility exists that the respondents who gave contrary responses misunderstood the survey item that led to the outlier responses, none of the contrary responses resulted from measurement error. Because none of the contrary responses distorted the research findings, all contrary responses were accepted for inclusion in the data analysis.

The outlier responses indicating that civilian employees at the Army organization perceived that their leaders use fair hiring and promoting procedures, but do not openly share those processes could signify that the small group of employees holds a high-level of trust for their leaders. The limited number of responses that resulted in this outlier data does not appear to be significant or generalizable to the population of the current research study.

The outlier responses indicating that civilian employees at the Army organization perceived that their leaders openly share hiring and promoting processes, but that those procedures are unfair could signify an underlying problem of procedure violations within the organization. Because of the unlikelihood that Army leaders openly share unfair hiring and promoting processes, the outlier responses might indicate that leaders at the Army organization are perceived by employees to employ hiring and promoting practices that are different from the openly shared hiring and promoting procedures. The limited number of responses that resulted in this outlier data does not appear to be significant or generalizable to the population of the current research study.

Implications

The implications of the current research study might improve the understanding of relationships between transparency, employee perceptions of fairness concerning organizational hiring and promoting practices, and employee morale. The implications derived from the results of the current research study and related to the research questions will be discussed in this section. The first research question inquired about the relationship between employee morale and the fairness perceptions of employees concerning organizational hiring and promoting practices. The second research question inquired about the relationship between increased transparency in the organization's hiring and promoting practices and employee fairness perceptions. The research results answered both research questions and suggested inferences that might lead to valuable changes in organizational leadership practices.

Research Question One Implications

The results of the data analysis indicated that a strong negative correlation exists between employee perceptions of the adverse affect of the hiring and promotion practices on employee morale and the fairness perceptions of employees concerning leaders' use of fair hiring and promoting practices. The results agree with the research results of Makawatsakul and Kleiner (2003) which declared that employee perceptions of leadership's use of unfair hiring and promoting processes may degrade employee morale. The results of the data analysis are significant because they indicate that improved employee perceptions might augment morale in organizations. The research question one findings may lead to changes in leadership practices aimed at fostering employee perceptions of fairness in an effort to nurture morale.

Research Question Two Implications

The results of the data analysis indicated that a positive correlation exist between employee perceptions of leaders' use of fair hiring procedures and employee perceptions of whether or not hiring procedures are openly shared. The analysis indicated that a positive correlation exist between employee perceptions of leaders' use of fair promoting procedures and employee perceptions of whether or not promoting procedures are openly shared. The findings revealed that 41.0% of the Federal employees surveyed agreed that increased transparency in hiring and promoting processes would improve their perception of fairness. These findings agree with research results from past scholars that supported the notion that employing transparency positively affects employee morale (Limas, 2005) and benefits organizations (Garten, 2002). These findings are significant because they may lead to changes in leadership practices that foster open communication, trust, and morale in organizations and reduce lawsuits relative to organizational hiring and promoting practices.

Recommendations

A survey released in 2007 included results supporting the belief that less than 60.0% of the employees at one Federal Government organization were satisfied with their jobs. The same survey led to the conclusion that only 67.5% of the employees, across the Federal Government, gave satisfactory reports concerning their jobs (Losey, 2007). The results of the current research study support the notion that negative employee perceptions relative to organizational leadership practices and the lack of organizational transparency might be related to the level of satisfaction employees report concerning their jobs. Several recommendations for improving employee perceptions of organizations are included in the following sections.

Recommendations for Leadership

One recommendation is that leaders should consider the results of the current research study when deciding the level of transparency to employ in organizational practices. Because the results of the current research study support the notion that increased transparency fosters employee perceptions of fairness and has a positive effect on organizational morale, leaders should share procedural information with employees to facilitate transparency and improve organizational productivity. Employing transparency to nurture perceptions of fairness and morale might assist in avoiding organizational debacles such as those that plagued mortgage industry giants in 2008.

Mortgage giants (i.e., Fannie Mae and Freddie Mac) suffered from failed leadership and their failures may prove to be costly to U.S. taxpayers. Kennedy (2008) reported that the U.S. Government must take control of the two mortgage giants in an attempt to support the collapsing housing market. The takeover is expected to cost U.S. taxpayers as much as $200 billion and hail as the biggest Government bailout in U. S. history (Kennedy, 2008). The Chairman of Freddie Mac took steps to enhance the organization's transparency (i.e., The Chairman registered the organization with the Securities and Exchange Commission) in an effort to improve the organization's health ("Freddie Mac to raise 5.5 billion dlrs in share offering", 2008).

Another recommendation is that leaders be a model for honesty, openness, and ethical behavior. Leaders should ask employees for feedback about organizational and leadership performance to gauge the effectiveness of any improvement practices employed. Employee feedback might be useful for facilitating procedural changes to help organizations and taxpayers avoid unnecessary financial burdens. The following sections include additional recommendations to foster organizational transparency, employee perceptions of fairness, and morale. A matrix of organizational improvements is included in Appendix G (see Table 10).

Transparency

Limas (2005) and Garten (2002) found that employing transparency positively affects employee morale and benefits organizations. The results of the current research study revealed that 41.0% of the Federal employees surveyed agreed that increased transparency in hiring and promoting processes would improve their perception of fairness. To improve organizational transparency, leaders should develop and openly share policies explaining internal controls governing the organization, share decision making criteria and results, and use committees to finalize major decisions.

Perceptions of Fairness

Historical research on fairness specified fairness criteria that included consistency, bias suppression, accuracy, ethicality, neutrality, and trust in the benevolence of authorities (Blader & Tyler, 2003). Because fairness perceptions are influenced by the satisfaction or violation of procedural rules (Gilliland, 1993), organizations should hold leaders who do not adhere to fair organizational procedures accountable (i.e., punish leaders who violate organizational procedural rules). Leaders should foster perceptions of fairness by using selection criteria to choose the most qualified applicants and allow employees to take part in organizational hiring and promoting processes. Research supports the belief that hiring and promoting procedures are perceived to be fairer when affected individuals have an opportunity to influence the process (Gilliland, 1993).

Morale

The results of the current research study revealed a correlation between employee perceptions of fairness concerning leaders' use of fair hiring and promoting practices and employee morale. The aforementioned actions taken to foster transparency and employee fairness perceptions will likely improve morale. Leaders should employ practices that positively influence leader-follower relationships (e.g., being honest, open, and ethical).

Leaders should refrain from abusing their power (e.g., using employee performance appraisals as a form of control).

Leaders are models, by virtue of their roles, and affect the behavior of their followers (Trevino, Hartman, & Brown, 2000). Models describe processes and are used to communicate values, attitudes, and behaviors (Trevino, Hartman, & Brown, 2000). The Ingram transparency-perception of fairness-morale model in Figure 5 illustrates the relationship between transparency, perception of fairness, and morale. In Figure 5, the variable "transparency" is directly controlled by leadership, but "perception of fairness", a variable that increases as the use of transparency is increased, is not directly controlled by leadership. The combination of transparency; employed by honest, open, and ethical leaders; and increasing perception of fairness work toward realizing a healthy level of morale in organizations.

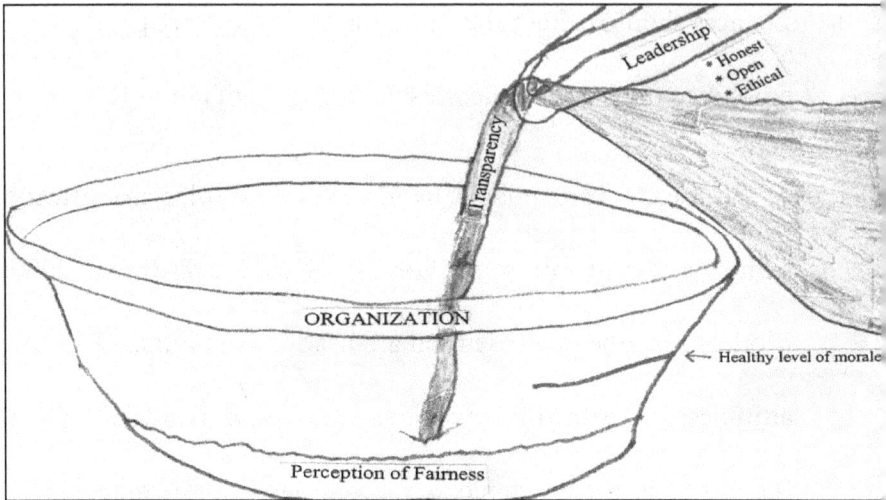

Figure 5. Ingram transparency-perception of fairness-morale

model

Researcher Reflections

This research was inspired by the experiences of the

researcher relative to perceived unfair hiring and promoting

activities while working for the U.S. DOD. Similar experiences

reported by relatives and friends of the researcher intensified the

researcher's desire to complete the current research study. The

experiences of the researcher resulted in bias and belief that the

survey respondents would overwhelmingly indicate perceptions

that leaders use unfair hiring and promoting practices, those practices adversely affect morale, and increased transparency would improve fairness perceptions. Although a large number of the respondents gave expected indications, many chose to remain non-committed and selected the neutral survey options (i.e., neither agree or disagree).

Future Research

Future research in fairness perceptions, morale, and organizational transparency could include a qualitative aspect (i.e., could employ a mixed-method design). Including a qualitative element would allow the researcher to seek explanations for responses given during the quantitative measurements. Future research could be carried out on a larger population (e.g., the U.S. DOD or the entire Federal Government). Increasing the population size would increase the significance of the results and might spur corrective actions if any widespread issues are found as a result of the research.

Future research could employ a survey instrument that seeks additional demographic data. The instrument could query respondents on their gender and ethnicity. The additional demographic data might illuminate race and gender-based issues and could lead to organizational and leadership interventions that might avert lawsuits and increased taxpayer burdens.

A future research opportunity could include employing regression analyses techniques. Regression analyses can be used to examine any variances found compared to the relationships supported by this research. Regression analyses can be used to test proposed hypotheses by establishing that a set of independent variables (e.g., transparency, procedural justice, and distributive justice) explains a portion of variances found in a dependent variable such as morale, motivation, or commitment (Bicknell, 2005). The current research study did not examine any direct relationship between transparency and morale. A regression analysis might support the establishment of a predictive relationship between the two variables.

The current research study minimally focused on distributive justice. Distributive justice concerns the fair and equitable distribution of resources such as salary and bonuses (Blader & Tyler, 2003). Past research examined the effects of distributive justice on employee commitment and productivity, but a gap in literature concerning any relationships between distributive justice and morale exists. Future research examining any relationship between distributive justice and morale might be valuable to leaders.

Conducting experimental research on diverse populations might yield useful results. The current research study focused on civilians employed at one Army organization in Florida. A future study might entail comparing results from multiple organizations. Future research could compare responses from respondents employed at a military organization to responses from respondents employed at a civilian (i.e., corporate) organization. The results of the research might clarify whether or not morale is affected differently in military organizations versus civilian organizations with respect to the amount of transparency employed by organizational leaders.

Summary

The current research study used a quantitative correlational research method to examine the relationships between organizational transparency, employee perceptions of fairness concerning organizational hiring and promoting practices, and employee morale. Sixty-four Federal employees were surveyed to address the research hypotheses and research questions posed in the current research study. The results of the data analyses of the responses given by the 64 subjects supported the hypotheses that a correlation exists between employee perceptions of fairness concerning leaders' use of fair hiring and promoting practices and employee morale and that increased transparency in organizational hiring and promoting processes greatly fosters employee perceptions of fairness.

The implications derived from the results of the current research study suggested that changes in leadership practices designed to foster employee fairness perceptions and improve morale might augment organizational relationships and productivity. Improved organizational transparency and morale might foster open communication and trust and reduce lawsuits relative to organizational hiring and promoting practices. Reduced organizational lawsuits might result in lower tax debt for the USA and lower tax burdens for U.S. citizens.

Future research suggested in the current research study included the addition of a qualitative aspect to the research method. Instrument improvements were also suggested to increase demographic data to identify any gender or race-based organizational issues. Increasing the population to improve the significance of the study results was suggested in chapter five.

Leaders should develop and openly share policies explaining internal controls governing the organization, take actions to foster employee perceptions of fairness, and employ practices that positively influence leader-follower relationships. These leadership actions, the Ingram transparency-perception of fairness-morale model, and the 13 nurturing leadership practices for organizational improvements proposed in the current research study (see Appendix G) might curtail any potential negative effects related to employee perceptions of unfairness. Lastly, the current research study recommended that leaders seek periodic feedback concerning organizational and leadership performance to gauge the effectiveness of any improvement practices employed to foster organizational transparency, employee perceptions of fairness, and morale.

REFERENCES

A white paper on research sampling. (2005). Retrieved June 20, 2007, from http://www.insightmas.com

Anonymous. (2008). Factors that determine morale among nurses in Queensland: The ability to cope with the rigours of the job is a key predictor of nurse morale. *Nursing Standard, 22(17), 18.*

Abdalla, H. F., Maghrabi, A. S., & Raggad, B. G. (1995). Assessing the perceptions of human resource managers toward nepotism: A cross-cultural study. *International Journal of Manpower, 19(8).*

Adams, J. S. (1965). Inequity in social exchange. *Advances in Experimental Social Psychology, 62, 335-343.*

Arvey, R. D., & Sackett, P. R. (1993). Fairness in selection: Current developments and perspectives. In N. Schmitt & W. C. Borman (Eds.), *Personnel selection in organizations,* San Francisco, CA: Jossey Bass.

Baehr, M. E. & Renck, R. (1958). The definition and measurement of employee morale. *Administrative Science Quarterly, 3(2), 157.*

Bagdadli, S., Roberson, Q., & Paoletti, F. (2006). The mediating role of procedural justice in response to promotion decisions. *Journal of Third World Studies, 23 (2), 83-102.*

Baldrige glossary. (2007). Retrieved July 29, 2007, from http://www.baldrige21.com/Baldrige%20Glossary%20HD.html

Banff Executive Leadership Inc. (2003, March). *Leadership acumen.* Retrieved April 9, 2008, from http://www.banffexeclead.com/NewsletterMar03.html

Bicknell, G. C. (2005). A secondary group level analysis of the effect of leader support on the relationship between combat exposure and post-combat aggression and violence. Doctoral dissertation, University of Texas at Austin, 2005.

Bies, R. J., & Shapiro, D. L. (1987). Interactional fairness judgments: The influence of causal accounts. *Social Justice Research 1(2), 199-218.*

Billikopf, G. E. (n. d.). Labor management in agriculture: Cultivating personnel productivity (2nd ed). Retrieved November 21, 2008, from http://www.cnr.berkeley.edu/ucce50/ag-labor/7labor/AgLabor.pdf

Blader, S. L. & Tyler, T. R. (2003, June). A four-component model of procedural justice: Defining the meaning of a "fair" process. *Personality and Social Psychology Bulletin, (29)6.*

Center for Creative Leadership. (1993). *What are morale, pride, and spirit? Building your team's morale, pride & spirit.* Retrieved July 23, 2007, from the EBSCO host research database.

Chan, A. (2005). Policy discourses and changing practice: Diversity and the university-college. *Higher Education, 50, 129–157.*

Church, A. H., & Waclawski, J. (1998). *Designing and using organizational surveys.* New York: Wiley.

Colquitt, J. A., Conlon, D. E., Wesson, M. J., Porter, C., & Ng, K. Y. (2001). Justice at the millennium: A meta-analytic review of 25 years of organizational justice research. *Journal of Applied Psychology, 86, 425-445.*

Creswell, J. W. (2002). *Educational research: Planning, conducting, and evaluating quantitative and qualitative research.* Columbus, OH: Merrill Prentice Hall.

Creswell, J. W. (2005). *Educational research: Planning, conducting, and evaluating quantitative and qualitative research.* Upper Saddle River, NJ: Pearson.

Cronbach, L.J. (1951). Coefficient alpha and the internal structure of tests. Psychometrika, 16, 297-334.

Cropanzano, R., Prehar, C. A., and Chen, P. Y. (2002). Using social exchange theory to distinguish procedural from interactional justice. *Group & Organization Management, 27 (3), 324-351.*

Denker, I. (2007). When and why did anti-corruption conventions come into being? Retrieved June 9, 2008, from http://www.transparency.org/global_priorities/international_ conventions/conventions_explained/history

Dixon, N. F. (1976). *On the psychology of military incompetence.* New York: Basic Books.

Dunn, S. (2000). *The significance of statistical significance.* Retrieved November 30, 2008, from http://cancerguide.org/significance.html

Federal Equal Employment Opportunity (EEO) laws prohibiting job discrimination. (n.d.). Retrieved November 21, 2008, from

http://carlislebarracks.carlisle.army.mil/installation/uploads
/files/Federal%20Laws%20Prohibiting%20Job%20Discrim
ination.doc

Fodchuk, K. M. & Sidebotham, E. J. (2005). Procedural justice in
the selection process: A review of research and suggestions
for practical applications. *The Psychologist-Manager
Journal, 8(2), 105–120.*

Freddie Mac to raise 5.5 billion dlrs in share offering. (2008).
Retrieved October 16, 2008, from
http://www.turkishpress.com/news.asp?id=242229&s=b&i=
&t=Freddie_Mac_to_raise_5.5_billion_dlrs_in_share_offeri
ng

Furnham, A. & Petrides, K. V. (2006). Deciding on promotions
and redundancies: Promoting people by ability, experience,
gender and motivation. *Journal of Managerial Psychology,
21(1), 6-18.*

Garten, J. E. (2002). Globalism without tears: A new social

compact for CEOs. Retrieved January 9, 2008, from

http://www.strategy-business.com/press/16635507/20863

Gay, L. R. (1996). *Educational research: Competencies for*

analysis and application (4th ed.). Beverly Hills, CA: Sage.

Gilliland, S. W. (1993). The perceived fairness of selection

systems: An organizational justice perspective. *Academy of*

Management Review, 18(4).

Glass, G. V., & Hopkins, K. D. (1996). Statistical methods in

education and psychology (3rd ed.). Needham Heights, MA:

Allyn & Bacon.

Golafshani, N. (2003). Understanding reliability and validity in

qualitative research. *The Qualitative Report, 8(4), 597-606.*

Retrieved February 22, 2008, from

http://www.nova.edu.sss/QR/QR8-4/golafshani.pdf

Government settles decades-old discrimination case for $508

   million. (2000, March). *Ethics Newsline*. Retrieved March

   11, 2008, from http://globalethics.org/newsline/

Grinberg, D. & Nazer, C. (2008). *Job bias charges rise 9% in*

   2007, EEOC reports. Retrieved March 10, 2008, from

   http://www.eeoc.gov/press/3-5-08.html

US Congress subpoenas documents on Abu Ghraib abuse

   [Electronic version]. (2006, July 7). Retrieved April 13,

   2008, from http://www.gulf-

   times.com/site/topics/article.asp?cu_no=2&item_no=96043

   &version=1&template_id=37&parent_id=17

Haun, D. E., Vivero, R., Leach, A., & Liuzza, G. (2002). Working

   short again? Absenteeism exacts toll from patient care,

   employee morale. Retrieved June 21, 2007, from

   http://www.mlo-online.com/articles/mlo0902short.htm

Hersey, P., Blanchard, K. H., & Johnson, D. E. (1996).

Management of organizational behavior: Utilizing human

resources. Upper Saddle River, NJ: Prentice-Hall.

Johnson, P. (2003). Thicker than water. *National review, 55(15).*

Retrieved June 18, 2007, from the Thomson Gale research

database.

Kahler, M. (2001). Leadership selection in the major multilaterals.

Retrieved January 23, 2008, from http://www.iie.com

Kaplan, D. M., & Ferris, G. R. (2001). Fairness perceptions of

employee promotion systems: A two-study investigation of

antecedents and mediators. *Journal of Applied Social*

Psychology, 31, 1204-1222.

Kennedy, H. (2008). Fannie Mae and Freddie Mac takeovers hits

home. Retrieved September 12, 2008, from

http://www.nydailynews.com/money/2008/09/07/2008-09-

07_fannie_mae_and_freddie_mac_takeovers_hit.html

Koppel, A., Barrett, T., & Tatton, A. (August 2006). Sen. Stevens is 'the secret senator'. Retrieved November 5, 2008, from http://www.cnn.com/2006/POLITICS/08/30/secret.senators/

Lainhart IV, J. W. (2000, July/August). Why IT governance is a top management issue. *Journal of Corporate Accounting & Finance, 11 (5), 33-40.*

Lambert, E. G., Cluse-Tolar, T., Pasupuleti, S., Hall, D. E., & Jenkins, M. (2005). The impact of distributive and procedural justice on social service workers. *Social Justice Research, 18(4), 411-427.*

Lemons, M. A., & Jones, C. A. (2001). Procedural justice in promotion decisions: Using perceptions of fairness to build employee commitment. *Journal of Managerial Psychology, 16 (4), 268-280.*

Limas, J. R. (2005). The influence of performance appraisal systems on leadership development. Doctoral dissertation, University of Phoenix, 2005.

Lindner, J. R. (1998). understanding employee motivation. *Journal of Extension, 36(3).* Retrieved November 21, 2008, from http://www.joe.org/joe/1998june/rb3.html

Losey, S. (2007). *DHS leaders aim to turn around poor morale.* Retrieved February 10, 2008, from http://www.federaltimes.com/index.php?S=3047128

Makawatsakul, N. & Kleiner, B. H. (2003). The effect of downsizing on morale and attrition. *Management Research News, 26(2-4), 52.*

Maze, R. (2007). *Many hiring managers snubbing vets.* Retrieved February 10, 2008, from http://www.federaltimes.com/index.php?S=3026827

McFadzean, F. & McFadzean, E. (2005). Riding the emotional roller-coaster: A framework for improving nursing morale. *Journal of Health Organization and Management, 19(4/5), 318-39.*

McKnight, D. H., Ahmad, S. and Schroeder, R. G. (2001). When do feedback, incentive control and autonomy improve morale? The importance of employee-management relationship closeness. *Journal of Managerial Issues, 13(4), 466-482.*

Messer, B. A. E., & White, F. A. (2006, Fall). Employees' mood, perceptions of fairness, and organizational citizenship behavior. *Journal of Business and Psychology, 21(1).*

Messmer, M. (2005, August). Avoiding today's top hiring mistakes. *Strategic Finance.* Retrieved May 30, 2006, from www.imanet.org/pdf/3243.pdf

Miceli, N. S. & Woods, D. R. (n.d.). *Resetting the reference points: Getting managers to use structured interviews.* Retrieved January 18, 2007, from http://owl.ben.edu/mamgt/Miceli12.doc

Neuman, W. L. (2003). *Social research methods: Qualitative and quantitative approaches*. Upper Saddle River, NJ: Prentice Hall.

Nunnally, J. C. (1978). *Psychometric theory.* New York: McGraw Hill.

Patton, M. Q. (2002). *Qualitative evaluation and research methods* (3rd ed.). Thousand Oaks, CA: Sage Publications, Inc.

Peelle, H. E. (2007). Reciprocating perceived organizational support through citizenship behaviors. *Journal of Managerial Issues*. Retrieved November 20, 2008, from *http://www.entrepreneur.com/tradejournals/article/print/17 3229709.html*

Public servants bend hiring rules, report finds. (2004, January 3). *The Record, A3.* Retrieved November 21, 2008, from Canadian Newsstand Torstar database.

Rajnandini, P., Schriesheim, C. A., & Williams, E. S. (1999, November-December). Fairness perceptions and trust as mediators for transformational and transactional leadership: A two-sample study. *Journal of Management, 25(6), 897.*

Rawlins, B. L. (2008, Spring). Measuring the relationship between organizational transparency and employee trust. *Public Relations Journal, 2(2).* Retrieved October 27, 2008, from http://www.prsa.org/prjournal/Vol2No2/Rawlins.pdf

Redpath, R., & Sheard, J. (2005). Domain knowledge to support understanding and treatment of outliers. Retrieved August 27, 2008, from www.csse.monash.edu.au/~rredpath/ICIA-REDPATH-DEC05.pdf

Richardson, J. E. (2004). *Annual editions: Business ethics (16th ed.).* Dubuque, IA: McGraw-Hill/Dushkin.

Roberts, A. (2004). A partial revolution: The diplomatic ethos and transparency in intergovernmental organizations. *Public Administration Review, (64)4.* Retrieved June 7, 2008, from http://www.aroberts.us/documents/journal/Roberts_PAR_R evolt_2004.pdf

Salkind, N. (2003). *Exploring research (5th ed.)*. New Jersey: Prentice Hall.

Sashkin, M., & Williams, R. L. (1990). Does fairness make a difference? *Organizational Dynamics, 9, 56-71.*

Sheldon, R. (2007). Market Value. *Utility Week, 27(1).* Retrieved April 14, 2008, from the EBSCOhost research database.

Singer, M. S., & Singer, A. E. (1991, October). Justice in preferential hiring. *Journal of Business Ethics, 10(10), 797.*

Somiah, T. (2006). Ethical dilemmas in Ghana: Factors that induce acceptance or rejection of bribes. Doctoral dissertation, University of Phoenix, 2006.

Sproull, N. D. (1995). *Handbook of research methods: A guide for practitioners and students in the social sciences (2nd. Ed.).* New Jersey: The Scarecrow Press.

Sproull, N. D. (2004). *Handbook of research methods: A guide for practitioners and students in the social sciences (3rd Ed.).* New Jersey: The Scarecrow Press.

Stephenson, C. (2004, January/February). Rebuilding trust: The integral role of leadership in fostering values, honesty and vision. *Ivey Business Journal Online.* Retrieved December 1, 2007, from the ProQuest research database.

Sterne, J. A. C., Smith, G. D. (2001). Sifting the evidence-what's wrong with significance tests? *British Medical Journal 322, 226–231.* Retrieved December 5, 2008, from http://www.pubmedcentral.nih.gov/articlerender.fcgi?tool=p ubmed&pubmedid=11159626

Stoddart, M. C. J. (2004). Generalizability and qualitative research in a postmodern world. *Graduate Journal of Social Science, 1(2).* Retrieved March 26, 2008, from http://www.gjss.nl/vol01/nr02/a05

Taris, T. W., Kalimo, R., & Schaufeli, W. B. (2002). Inequity at work: Its measure and association with worker health. *Work & Stress, 16(4), 287-301.*

Transparency International. (2007a). *Addressing corruption and building integrity in defense establishments.* Retrieved January 23, 2008, from http://www.transparency.org/publications/publications/working_paper_no_2_defence

Transparency International. (2007b). When and why did anti-corruption conventions come into being? Retrieved April 9, 2008, from http://www.transparency.org/global_priorities/international_conventions/conventions_explained/history

Trevino, L.K, Hartman, L.P. & Brown, M. (2000). Moral person and moral manager: How executives develop a reputation for ethical leadership. *California Management Review, 42(4), 128-147.*

Truxillo, D. M., Bauer, T. N., Paronto, M. E., & Campion, M. A. (2002). Selection fairness information and applicant reactions: A longitudinal field study. *Journal of Applied Psychology, 87, 1020–1031.*

Tyler, T. R., and Bies, R. J. (1990). Beyond formal procedures: The interpersonal context of procedural justice. In J. S. Carroll (Ed.), *Applied Social Psychology in Business Settings, 77-98.*

Walker, K. & Pagano, B. (2005). *Transparency: The clear path to leadership credibility.* Retrieved January 23, 2007, from linkageinc.com/company/.../01_05_Article_Transparency_WalkerPagano.pdf

Ziomek, J. (2005). *Journalism, transparency and the public trust: A report of the eighth annual Aspen Institute conference on journalism and society.* Retrieved October 27, 2008, from http://www.aspeninstitute.org/atf/cf/{deb6f227-659b-4ec8-8f84-8df23ca704f5}/jourtransptext.pdf

APPENDICES

APPENDIX A: RESEARCH METHODOLOGY MAP

Research Methodology Map

1. PROBLEM DEFINITION AND APPROACH
Identify problem, Identify need for the study, Introduce research methodology, Identify the study population, Identify the research variables, Identify the research questions, Identify the hypotheses, and Identify the study's theoretical framework.

2. LITERATURE REVIEW
Review existing literature concerning the research topic, Provide a historical overview of the topic being studied, Identify the broad research area under which the topic falls, Outline the current research findings relative to the research topic being studied, and Prepare plans for data collection and analysis.

3. RESEARCH METHODOLOGY
Identify a validated survey instrument that will be used for the study, Outline procedures for sampling, and Outline the data collection and analysis process.

4. PROPOSAL APPROVAL
Work closely with research committee to finalize the research proposal (i.e., the first three chapters of the research study), submit proposal to the Institutional Review Board (IRB) for approval, and obtain proposal approval from the IRB.

5. DATA COLLECTION AND ANALYSIS
Effect the data collection process (i.e., administer the survey and collect responses), Validate survey data, Formulate the data for analysis, Analyze the survey data using the Statistical Package for the Social Sciences (SPSS) analysis software tool, and formulate data analysis results.

6. CONCLUSIONS AND RECOMMENDATIONS
Provide research-supported opinions, commentary, judgments, conclusions, and observations about the research study; Highlight the importance, significance, and meaning of the inquiry; and Suggest areas for further research.

APPENDIX B: INFORMED CONSENT

This multiple-choice 10 question survey is designed to examine the relationship between organizational transparency, employee morale, and employee fairness perceptions concerning organizational hiring and promoting practices at your place of employment. Your responses will remain anonymous and confidential and will be used in a final report outlining any relationships found. Any electronic data collected will be deleted after being secured for three years. Since respondent identification information will not be collected, there is no risk that your answers will be connected to you in any way. Your responses may have a positive impact on the hiring and promoting processes at your place of employment. You must be at least 18 years of age to participate in this survey. Selecting the oval shown below beside the statement that gives consent to share information in the current research study is equivalent to giving a signature-based consent. Selecting the first oval below assures the following:

I acknowledge that I understand the nature of the study, the potential risks to me as a participant, and the means by which my identity will be kept confidential. My signature on this form also indicates that I am 18 years old or older and that I give my permission to voluntarily serve as a participant in the study described.

O I understand the above statements and give consent for my information to be used in the study

O I understand the above statements and do NOT give consent for my information to be used in the study

APPENDIX C: SURVEY INTRODUCTORY LETTER

Dear Respondent,

Please consider completing the short 10-question multiple-choice electronic survey referenced below. The survey is designed to assist in examining the relationship between organizational transparency, employee morale, and employee fairness perceptions concerning organizational hiring and promoting practices at your place of employment. Your responses will remain anonymous and confidential. Any electronic data collected will be deleted after being secured for three years. Since respondent identification information will not be collected, there is no risk that your answers will be connected to you in any way. The answers you give will be used in a final report outlining any relationships found. Your responses may have a positive impact on the hiring and promoting processes at your place of employment. You must be at least 18 years of age to participate in this survey. If you select the first oval below, you will be signing this form and giving your consent to take part in the current research study. Selecting the first oval below assures the following:

I acknowledge that I understand the nature of the study, the potential risks to me as a participant, and the means by which my identity will be kept confidential. My signature on this form also indicates that I am 18 years old or older and that I give my permission to voluntarily serve as a participant in the study described.

O I understand the above statements and give consent for my information to be used in the study

O I understand the above statements and do NOT give consent for my information to be used in the study

APPENDIX D: EMPLOYEE PERCEPTION-EMPLOYEE MORALE SURVEY COVER LETTER AND INSTRUMENT

Dear Respondent:

Many researchers have questioned if organizational morale losses might be related to fairness perceptions concerning organizational hiring and promoting procedures. Researchers have questioned if an organization's sharing of information (promoting transparency) relative to hiring and promoting decisions influence perceptions of fairness. This survey is designed to examine the relationship between organizational transparency, employee morale, and employee fairness perceptions concerning organizational hiring and promoting practices at your place of employment. If you would please take a few moments to answer (on your own time and away from your place of employment) and return this 10 question survey, your answers might be instrumental in helping your leaders design future processes to positively affect organizational morale.

Your responses will remain anonymous and confidential and will be used in a final report outlining any relationships found between organizational transparency, employee morale, and employee fairness perceptions concerning organizational hiring and promoting practices. I assure you that your name will not be included in any report documenting the results of this research. Thank you in advance for completing this survey.

	Strongly disagree	Disagree	Neither agree or disagree	Agree	Strongly agree
1. It is my perception that leaders at my organization use fair procedures in hiring decisions.					
2. It is my perception that leaders at my organization use fair procedures in promotion decisions.					
3. It is my perception that job openings are well communicated to potential new hire candidates.					
4. It is my perception that qualification requirements for job openings are written fairly and accurately for the positions.					
5. It is my perception that promotion opportunities are well communicated through formal channels to potential candidates.					

6. It is my perception that the best candidate for the job is usually hired, regardless of gender, race, or age.					
7. Current hiring practices adversely affect my morale on the job.					
8. Current promotion practices adversely affect my morale on the job.					
9. It is my perception that hiring and promoting processes are openly shared with employees.					
10. If explanations of the hiring and promoting processes were more openly shared with employees, my perception of fairness would improve.					

APPENDIX E: EMPLOYEE RESPONSES TO SURVEY ITEMS

Table 8

onses (R1 through R64) to Survey Items (SI-1 through SI-10)

	SI-1	SI-2	SI-3	SI-4	SI-5	SI-6	SI-7	SI-8	SI-9
R1	5.00	5.00	4.00	5.00	5.00	3.00	2.00	2.00	4.00
R2	3.00	3.00	1.00	3.00	3.00	2.00	3.00	3.00	2.00
R3	2.00	2.00	4.00	4.00	2.00	4.00	2.00	3.00	2.00
R4	5.00	5.00	4.00	3.00	3.00	4.00	2.00	2.00	4.00
R5	2.00	2.00	3.00	3.00	3.00	2.00	3.00	4.00	2.00
R6	1.00	1.00	1.00	1.00	1.00	1.00	5.00	5.00	1.00
R7	3.00	2.00	3.00	4.00	3.00	2.00	3.00	4.00	3.00
R8	4.00	4.00	3.00	4.00	3.00	4.00	2.00	3.00	3.00
R9	2.00	1.00	2.00	2.00	2.00	1.00	5.00	5.00	1.00
R10	2.00	2.00	4.00	4.00	4.00	2.00	4.00	4.00	4.00
R11	4.00	4.00	4.00	4.00	4.00	4.00	3.00	3.00	3.00
R12	4.00	4.00	3.00	4.00	3.00	3.00	2.00	3.00	3.00
R13	4.00	4.00	5.00	5.00	5.00	4.00	3.00	3.00	4.00
R14	2.00	2.00	3.00	3.00	2.00	1.00	5.00	5.00	4.00
R15	3.00	3.00	4.00	3.00	4.00	4.00	2.00	2.00	3.00
R16	3.00	3.00	2.00	2.00	3.00	3.00	3.00	3.00	2.00
R17	4.00	4.00	3.00	4.00	3.00	2.00	1.00	1.00	4.00
R18	3.00	3.00	3.00	3.00	3.00	2.00	3.00	3.00	3.00
R19	4.00	4.00	3.00	3.00	3.00	4.00	2.00	3.00	2.00
R20	2.00	2.00	2.00	2.00	2.00	2.00	5.00	5.00	2.00
R21	4.00	4.00	4.00	4.00	4.00	2.00	5.00	2.00	4.00

R22	1.00	1.00	2.00	3.00	2.00	2.00	4.00	4.00	3.00
R23	3.00	3.00	2.00	3.00	2.00	2.00	3.00	3.00	3.00
R24	5.00	1.00	3.00	2.00	4.00	1.00	1.00	3.00	4.00
R25	2.00	1.00	4.00	4.00	4.00	1.00	4.00	5.00	1.00
R26	4.00	4.00	3.00	3.00	3.00	3.00	2.00	2.00	4.00
R27	5.00	5.00	4.00	5.00	5.00	5.00	2.00	2.00	5.00
R28	2.00	2.00	5.00	4.00	4.00	1.00	5.00	4.00	2.00
R29	2.00	2.00	2.00	2.00	2.00	2.00	3.00	3.00	2.00
R30	2.00	2.00	2.00	3.00	2.00	2.00	4.00	4.00	2.00
R31	5.00	5.00	5.00	4.00	5.00	5.00	1.00	1.00	5.00
R32	4.00	4.00	3.00	4.00	3.00	3.00	2.00	3.00	3.00
R33	3.00	3.00	4.00	3.00	3.00	3.00	3.00	3.00	3.00
R34	4.00	4.00	5.00	5.00	5.00	4.00	3.00	3.00	4.00
R35	2.00	2.00	3.00	3.00	2.00	1.00	5.00	5.00	4.00
R36	2.00	2.00	4.00	4.00	2.00	4.00	2.00	3.00	1.00
R37	5.00	5.00	4.00	5.00	5.00	3.00	2.00	2.00	4.00
R38	5.00	5.00	4.00	3.00	3.00	4.00	2.00	2.00	4.00
R39	4.00	4.00	3.00	4.00	3.00	2.00	1.00	1.00	4.00
R40	1.00	1.00	1.00	1.00	1.00	1.00	5.00	5.00	1.00
R41	3.00	2.00	3.00	4.00	3.00	2.00	3.00	4.00	3.00
R42	4.00	4.00	3.00	4.00	3.00	4.00	2.00	3.00	3.00
R43	3.00	3.00	1.00	3.00	3.00	2.00	3.00	3.00	2.00

R44	2.00	2.00	4.00	4.00	4.00	2.00	4.00	4
R45	4.00	4.00	4.00	4.00	4.00	4.00	3.00	3
R46	3.00	3.00	4.00	3.00	3.00	3.00	3.00	3
R47	3.00	3.00	4.00	3.00	4.00	4.00	2.00	2
R48	3.00	3.00	2.00	2.00	3.00	3.00	3.00	3
R49	3.00	3.00	3.00	3.00	3.00	2.00	3.00	3
R50	2.00	1.00	4.00	4.00	4.00	1.00	4.00	5
R51	2.00	2.00	2.00	2.00	2.00	2.00	5.00	5
R52	5.00	5.00	4.00	5.00	5.00	5.00	2.00	2
R53	1.00	1.00	2.00	3.00	2.00	2.00	4.00	4
R54	4.00	4.00	3.00	3.00	3.00	4.00	2.00	3
R55	5.00	1.00	3.00	2.00	4.00	1.00	1.00	3
R56	4.00	4.00	4.00	4.00	4.00	2.00	5.00	2
R57	4.00	4.00	3.00	3.00	3.00	3.00	2.00	2
R58	2.00	2.00	3.00	3.00	3.00	2.00	3.00	4
R59	2.00	2.00	5.00	4.00	4.00	1.00	5.00	4
R60	2.00	2.00	2.00	2.00	2.00	2.00	3.00	3
R61	2.00	1.00	2.00	2.00	2.00	1.00	5.00	5
R62	2.00	2.00	2.00	3.00	2.00	2.00	4.00	4
R63	5.00	5.00	5.00	4.00	5.00	5.00	1.00	1
R64	3.00	3.00	2.00	3.00	2.00	2.00	3.00	3

APPENDIX F: DESCRIPTIVE STATISTICAL DATA

Table 9

Descriptive Statistical Data Relative to Respondents' Survey Answers

Descriptive Statistics

Survey items	Number of responses	Number of valid responses	Mean	Mode
SI-1	64	64	3	2
SI-2	64	64	3	3
SI-3	64	64	3	2
SI-4	64	64	3	3 and 4
SI-5	64	64	3	3
SI-6	64	64	3	2
SI-7	64	64	3	3
SI-8	64	64	3	3
SI-9	64	64	3	2, 3, and 4
SI-10	64	64	3	3

APPENDIX G: LEADERSHIP IMPROVEMENTS MATRIX

Table 10

eadership Improvements

	Leadership Improvements
Concept	Nurturing leadership practices
Transparency	Openly and honestly share policies
	Share decision making criteria and results
	Use committees to finalize major decisions
Perceptions of fairness	Foster transparency
	Suppress biases
	Promote ethical behavior
	Adhere to organizational procedures
	Penalize procedure violators

www.ingramcontent.com/pod-product-compliance
Lightning Source LLC
Chambersburg PA
CBHW060335030426
42336CB00011B/1358